TIME, SPACE AND PHILOSOPHY

In this book, Christopher Ray addresses fundamental issues in the philosophy of space and time, while avoiding daunting technicalities and jargon.

Always careful to elucidate the philosophical problems associated with space and time, Ray examines the work of Newton, Einstein, Hawking and other scientific giants and discusses the reactions of philosophers to this work – from metaphysical worries about the nature and reality of space and time to questions about the status of relativity and its rival theories.

He investigates the puzzling nature of space – from the infinitesimally small to the unimaginably large, the disturbing paradoxes of time and time travel, and the curious ideas of modern cosmology – from the big bang and the possibility of creation *ex nihilo* to the quantum world of black holes.

Christopher Ray is Assistant Professor in History and Philosophy of Science at Portland State University, Oregon. His published works include *The Evolution of Relativity* (1987).

PHILOSOPHICAL ISSUES IN SCIENCE

General Editor
W.H. Newton-Smith

THE RATIONAL AND THE SOCIAL
J.R. Brown

THE NATURE OF DISEASE
Lawrie Reznek

INFERENCE TO THE BEST EXPLANATION
Peter Lipton

THE PHILOSOPHICAL DEFENCE OF PSYCHIATRY
Lawrie Reznek

MATHEMATICS AND THE IMAGE OF REASON
Mary Tiles

THE LABORATORY OF THE MIND
J.R. Brown

Forthcoming

METAPHYSICS OF CONSCIOUSNESS
William Seager

TIME, SPACE AND PHILOSOPHY

Christopher Ray

London and New York

First published 1991
by Routledge
11 New Fetter Lane, London EC4P 4EE

Simultaneously published in the USA and Canada
by Routledge
29 West 35th Street, New York, NY 10001

Reprinted 1992

© 1991 Christopher Ray

Typeset in 10/12pt Baskerville, Linotron 202 by Columns, Reading
Printed and bound in Great Britain by
Biddles Ltd, Guildford and King's Lynn

All rights reserved. No part of this book may be reprinted or reproduced or utilized in any form or by any electronic, mechanical, or other means, now known or hereafter invented, including photocopying and recording, or in any information storage or retrieval system, without permission in writing from the publishers.

British Library Cataloguing in Publication Data
Ray, Christopher
Time, space and philosophy. – (Philosophical issues in science).
1. Space & time. Philosophical perspectives
I. Title II. Series
115

Library of Congress Cataloging in Publication Data
Ray, Christopher.
Time, space, and philosophy / Christopher Ray.
p. cm. – (Philosophical issues in science)
Includes bibliographical references and index.
1. Space and time. I. Title. II. Series.
BD632.R39   1991   90–24118
115—dc20

ISBN 0–415–03221–0   ISBN 0–415–03222–9 (pbk.)

For Carol
χρόνῳ ποτέ

CONTENTS

Preface   x
Introduction   1

1 ZENO AND THE LIMITS OF SPACE
   AND TIME   5
   Introduction   5
   Divisibility versus indivisibility   6
   Infinitesimals and limits   11
   Thomson's infinite super-task   14
   The parallel task paradox   15
   Abstractions and the physical world   20

2 CLOCKS, GEOMETRY AND
   RELATIVITY   24
   Introduction   24
   My time and your time   33
   The paradox of the twins: for ever young?   36
   From twins to triplets   41
   Phantoms of perspective   44

3 TRAVELLING LIGHT   46
   Introduction   46
   Measuring the speed of light   49
   Absolute simultaneity?   53
   Slow clock transport   57
   Spacelike travel: a tale of two tachyons   60
   Just the two of us: across the universe?   66

4 A CONVENTIONAL WORLD?   69
   Introduction   69
   When parallel lines meet   71

CONTENTS

Will the real geometry please stand up?	74
Convention and topology	79
Dimensions	82
The future of the universe	84
The Cosmological Principle: convention or fact?	86
The underdetermination of theory by data	90

5 NEWTON AND THE REALITY OF SPACE AND TIME — 99
- Introduction — 99
- Absolute space and time — 100
- Matter in the Newtonian world — 103
- Leibniz and relationism — 105
- Clarke's defence of Newton — 108
- Absolute motion without absolute space? — 113

6 MACH AND THE MATERIAL WORLD — 116
- Introduction — 116
- Mach's relationism — 118
- Simplicity and science — 120
- Positivism in action — 122
- Can we see space? — 125
- Experiment and intervention — 127

7 EINSTEIN AND ABSOLUTE SPACETIME — 131
- Introduction — 131
- Mach's Principle — 133
- Absolutely, Professor Einstein? — 134
- Empty, almost empty, and rotating worlds — 139
- Relationism and relativity: an empirical view? — 143
- The hole argument and spacetime points — 146

8 TIME TRAVEL — 151
- Introduction — 151
- Spacetime structure — 154
- Back to the past — 156
- Forward to the past — 166
- Correlations and backwards causation — 171

9 EINSTEIN'S GREATEST MISTAKE? — 176
- Introduction — 176
- Space and infinity — 177
- Einstein's universe — 180

CONTENTS

The cosmological constant: did Einstein blunder?	184
Laws and theoretical change	187
The Anthropic Principle	189
10 COSMOLOGICAL CONUNDRUMS	**193**
Introduction	193
The big bang: a singular idea	196
The beginning of time?	199
Inflationary cosmology: something for nothing?	204
Black holes	209
Cosmic censorship	211
Determinism versus indeterminism	215
CONCLUSION: RELATIVITY – JUST ANOTHER BRICK IN THE WALL?	**217**
Introduction	217
What is a theory?	218
The structure and scope of spacetime theories	221
The last word?	226
NOTES	**229**
SELECT BIBLIOGRAPHY	**260**
INDEX	**263**

PREFACE

This book presents my reflections upon a series of problems about time and space. Much discussed here has a long and distinguished heritage. I have every reason for gratitude to both earlier and present generations of scientists and philosophers for their exploration and clarification of our ideas about space and time: from Samuel Clarke's defence of Newton to Hans Reichenbach's empiricism; from Aristotle's discussion of Zeno on motion to Hugh Mellor's thoughts about time and time travel; from Albert Einstein's revolutionary thoughts about matter and the source of inertia to Stephen Hawking's equally startling discussion of the properties of black holes. I also have every reason to thank those who helped, in various ways, with this book, reading part or all of the various drafts or discussing the ideas involved, and giving so many valuable suggestions, and steering me away from error too often for me to have anything else but a marked sense of my fallibility. I am especially grateful to Harvey Brown and Bill Newton-Smith of Oxford University, Marthe Chandler at DePauw University, Carl Hoefer of Stanford University, Alexander Rueger at the University of Oregon and Robert Weingard of Rutgers University. I am grateful too for the assistance given to me by Richard Stoneman and the editorial staff at Routledge, and for their patience. And Carol Ray, reading the manuscript as a non-specialist, did more than anyone to help me to clarify those ideas which were expressed too clumsily or too technically. So the merits of this book derive in part from the endeavours of others; but the defects you must blame on me.

Some of the material in this book is based on articles published in journals, with revisions where appropriate, and I am grateful to the editors of the journals involved for allowing me to use this

PREFACE

material here. The central part of Chapter 1 appears as 'Paradoxical tasks' in *Analysis* **50** 2 (1990); the last section of Chapter 3 is based on 'Can we travel faster than light?' in *Analysis* **42** 1 (1982); the final section of Chapter 7 is based on part of a review, written together with Carl Hoefer, of John Earman's *World Enough and Space-Time*, in the *British Journal for the Philosophy of Science* **42** 3 (1991); and much of Chapter 9 appeared in 'The cosmological constant: Einstein's greatest mistake?' *Studies in the History and Philosophy of Science* **21** 4 (1990).

My thanks must go as well to Mr P, never far from any centre of activity, for his slumbering and I hope appreciative feline reflections on my endeavours.

Christopher Ray
Portland, Oregon, USA

INTRODUCTION

Under the startling headlines 'Revolution in science: New theory of the Universe: Newtonian ideas overthrown', the *New York Times* reported, in 1919, the effects of Sir Arthur Eddington's dramatic confirmation of Einstein's General Theory of Relativity and its prediction that a light ray from a distant star would 'bend' in the curved space close to the Sun:

> Yesterday afternoon in the rooms of the Royal Society, at a joint session of the Royal and Astronomical Societies, the results obtained by British observers of the total solar eclipse of May 29 were discussed. The greatest possible interest had been aroused in scientific circles by the hope that rival theories of a fundamental physical problem would be put to the test, and there was a very large attendance of astronomers and physicists. It was generally accepted that the observations were decisive in verifying the prediction of the famous physicist, Einstein, stated by the President of the Royal Society as the most remarkable scientific event since the discovery of the planet Neptune. But there was a difference of opinion as to whether science had to face merely a new and unexplained fact, or to reckon with a theory that would completely revolutionize the accepted fundamentals of physics.
>
> *(New York Times* 1919)[1]

Later that year, Einstein was invited to explain his ideas to the British public. In a short article, he presented the essential features of his theory: he told the readers of *The Times* that 'In the generalised theory of relativity, the doctrine of space and time ...

is no longer one of the absolute foundations of general physics' (Einstein: 28 November 1919).

Our concepts of space and time, already challenged by Einstein's Special Theory of Relativity, were now under further attack from his General Theory. Few understood the implications of Einstein's work in those early years. Many found it hard to break free from the well-established Newtonian ideas. But more and more the scientific community embraced Einstein's theories. Some of the initial implications of both theories were hard to swallow: the idea that time is not an absolute framework; and the possibility of a non-Euclidean universe in which the three internal angles of a triangle do not add up to 180 degrees. Even Einstein found some of the implications hard to stomach: his equations were consistent with the possibility of an expanding universe – a possibility which he initially rejected in a move which he came to regard as his greatest mistake. And more surprises were to come as the theories were developed further: the big bang, time travel, and black holes all seemed to be consistent with the ideas of relativity theory.

In this book, we shall explore some of the major ideas and problems behind our views of space and time. Most of the central questions about space and time arise from considering the ideas of scientists such as Isaac Newton, Ernst Mach, Albert Einstein, and Stephen Hawking. So we must consider the essential features of the work on space and time by such scientists as these: from speculations about how many dimensions space might have to the problem of infinitesimals; from questions about whether space and time are infinite to worries about the scientific status of entities which cannot be seen; from the ideas of black holes and the big bang to conjectures about time travel. We shall then be in a better position to understand the philosophical issues connected with all these problems.

In Chapter 1, we shall look at the five paradoxes presented by the early Greek philosopher Zeno. His worries about the way we regard space, time, and motion have a clear message for the way we think of geometry and its applicability to the physical world. The problems of geometry are pursued further in Chapters 2 to 4. First we shall discuss the celebrated paradox of the twins and introduce the less well-known paradox of the triplets; we shall then investigate the importance of the speed of light in relativity theory, asking, amongst other questions, what happens when we relax the generally held convention that nothing travels faster than light;

INTRODUCTION

and then we shall focus on the general implications of relativity's commitment to non-Euclidean geometries. In Chapters 5, 6, and 7, we shall look closely at the question of absolute and relational space and time, first through the arguments of Newton and Leibniz, and then through the ideas of Mach and Einstein. We shall see that the problems identified by Newton may be raised in both Newtonian and relativistic contexts. Chapter 8 focuses on the problems and possibilities of time travel. We shall discuss several ways in which time travel might be possible; but we shall find that some of them may involve logical contradictions or may require rather peculiar views of the physical world. The problems of classical and modern ideas of cosmology are addressed in Chapters 9 and 10. Particular attention is given to the cosmological constant – the idea dismissed by Einstein as a blunder. But we shall also review problems connected with black holes and the big bang. The final chapter presents an overall impression of the status of claims about space, time, and motion: how much should we believe of the stories told to us by physicists when they seem to change their minds so often?

Throughout I have aimed to draw a balance between explaining the physics and examining the philosophical assumptions, arguments, and perspectives involved in the various physical accounts ahead. I have tried to keep technical details to a minimum, but sometimes the problems which we meet cannot be grasped without at least some appreciation of the mathematical and geometrical ideas involved. Where possible, I have used diagrams to help the reader visualise the situations being discussed. In writing this book, I have tried to provide a comprehensive, up-to-date, and accessible introduction to the philosophy of space and time, to help those without specialist backgrounds in the physics of space and time begin to understand (and not just be dazzled by) some of the fundamental issues arising from classical and modern ideas of space and time – issues which will also introduce the reader to philosophical problems in metaphysics, the theory of knowledge, the philosophy of religion, and the philosophy of science. However, I hope that many readers will regard this book as a starting-point for further studies in the philosophy of space and time. So a select bibliography reviews the most important and helpful literature in the field. And detailed notes to each chapter amplify the text, suggest further reading, and point those wishing to engage in further research in the right direction.

The ideas of space and time provide us with a rich and rewarding field of study. The challenge which faced Newton and Einstein may be shared by everyone. We may not have their genius, but we can share their insights. And these insights can give us a better appreciation of the role of philosophy as it meets the problems of science.

1
ZENO AND THE LIMITS OF SPACE AND TIME

INTRODUCTION

We typically think of space and time as three dimensions plus one. Mathematicians tell us that each dimension may be continuously sub-divided. But they also tell us that we may construct model universes with rather different properties. We may have other structures which may not be continuously sub-divided. And, to complicate matters, we may construct worlds with whatever dimensionality we please. So, can we really chop 'real' space and time up as small as we like?

The pre-Socratic philosopher Zeno of Elea – a Greek settlement in Southern Italy – is said to be responsible for five 'paradoxes' which wrestle with the properties of space, time, and motion. The main focus of Zeno's paradoxes is the 'small-scale' character of space and time. Is this small-scale structure really continuous, or is it 'indivisibly atomistic' or 'discrete' in some sense? If three-dimensional space is a continuum, then we may continuously and indefinitely sub-divide its parts. But if space or time are discrete in some way, then any process of sub-division will have a definite limit. Aristotle gives a brief and perhaps incomplete account of the first four paradoxes in his *Physics*, and Simplicius discusses the fifth in his commentary on Aristotle.[1] Zeno is thought to have produced his ideas around 460 BC. We shall review Zeno's discussion, and we shall find that these paradoxes do identify some real difficulties for our 'continuum' view of space and time.

Many mathematicians and philosophers believe that a thorough acquaintance with the mathematics of the continuum should be sufficient to dispel any worries that might arise from Zeno's paradoxes. But the problems raised by Zeno live on and some writers, including the philosopher Wesley Salmon and the theoretical physicist Roger Penrose, advise against any uncritical and complete acceptance of the role of the continuum in our physical theories.[2] A related problem, suggested by James Thomson in 1954, concerns the paradoxical nature of any super-task consisting of an infinite number of tasks. I shall argue that this problem is genuinely paradoxical, on the mathematicians' own terms. But I shall not join Zeno in rejecting the reality of a complex, diverse world. I shall merely question the extent to which mathematics and geometry may serve as an adequate model for the physical world.

Imagine that we have two theories about the way objects move in the world. One theory assumes that space and time may be continuously sub-divided. The other denies this. But also imagine that both theories are perfectly consistent with every measurement and observation we can possibly make. If we can actually construct such an empirically impeccable rival to the 'continuum' theory, then we might begin to wonder about the status of the continuum. We may be willing to admit that it gives us an extremely useful way of organising our experience. But should we believe that the world is really like that? The advantage of mathematics is that it helps us to think clearly about those structures which we believe to be the actual structures of the world; but the problem with mathematics is that it allows us to generate all sorts of weird and wonderful possible structures for the world. The job of sorting out which, if any, we should accept as the 'real' picture is left to the physicist. And sometimes the choice is far from straightforward.

DIVISIBILITY VERSUS INDIVISIBILITY

Zeno's paradoxes of space, time, and motion attack the very idea of the divisibility of space and time. We begin by imagining a distance or a temporal duration which is divided by two; and we imagine that the process of division is continued. Why may we not imagine that the process could continue indefinitely? Zeno tells us that any assumption that the process could go on indefinitely will lead us into logical contradictions. But he also argues that any assumption that the process has some definite limit also leads us into just as much trouble. The first four paradoxes reveal the dilemma:

ZENO AND THE LIMITS OF SPACE AND TIME

1 Achilles and the tortoise
Zeno asks us to imagine a race between Achilles and a tortoise in which the tortoise is allowed to start first. After an agreed time, Achilles sets off in pursuit. Although it seems entirely obvious that the race is a mis-match and that Achilles will all too soon overtake the tortoise, Zeno raises a doubt in our minds. For in order to overtake the tortoise Achilles must first reach the point where the tortoise was when Achilles was given the signal to start in pursuit. Let us call this first point P. But when he reaches point P, the tortoise will now be a little further on at point Q. Achilles now must reach Q if he is to catch the tortoise. Yet when he arrives at Q the tortoise is still ahead at R. When Achilles gets to R, the tortoise has reached S. The race continues just like this: every time Achilles reaches the tortoise's last 'staging-post' the tortoise has moved further on to a new post. Of course, the distance between the two gets shorter and shorter all the time. But Achilles is always behind! So despite first appearances Achilles cannot even catch let alone overtake the tortoise.

2 The racecourse (or dichotomy paradox)
Here Zeno not only argues that an athlete would never finish, say, a 100-metre race, it also seems that the athlete could not even get started! To reach the end of the track, the athlete would first have to reach the 50-metre point. Having run 50 metres, the athlete would now have to reach the half-way point between the 50-metre point and the finish line. That would take the athlete to the 75-metre mark. But now the athlete would have to reach the half-way point between this mark and the finish. No matter how far the athlete gets down the track, there would always be yet another 'half-way' point to reach between the point where the athlete is and the finishing line. So the athlete would get closer and closer to the end of the track, but would never actually reach the finish. For there would be an infinite number of half-way points ahead of the athlete. This might seem bad, but an associated argument implies that the race would not even begin. For to reach the finishing line demands that the athlete would first need to reach the 50-metre mark; and to reach the 50-metre mark demands that the athlete would already have reached the 25-metre point; and to reach that point would require that athlete to have got to the 12.5-metre mark; and so on. As we keep dividing the distance by two, we get closer to the starting

line, but we never actually reach it. And we may divide these distances an infinite number of times. So to reach the end of the track there would be an infinite number of distances to run through. Indeed, no matter how short the track, there would always be an infinite number of distances ahead. The athlete would be stuck at the start. To go any distance at all, the athlete would have to run through an infinite number of distances – and how could that be possible?

3 The arrow

Take a high-speed photograph of an arrow in flight and you may find it hard to disagree with Zeno's assertion that such an arrow occupies exactly that space which is equal to its own shape and size. We seem to have captured the arrow at an instant of time. At such an instant the arrow is motionless. If it were not motionless, the instant of time could be sub-divided: now the arrow is here, now there. Yet the entire flight of the arrow could be captured in a series of instantaneous photographs. At every instant, the arrow is motionless. There is no time between the instants for the arrow to move on to the next instant. For such a time would be composed of instants itself. So how can an always motionless object move?

4 The moving rows (or the stadium)

Imagine a stadium in which a column of soldiers passes a column of soldiers at attention so that each step brings every soldier in the moving column into line with the next comrade in the stationary column; a third column of soldiers is also moving, but in the opposite direction, so that with each step the soldiers here also are brought into line with the next comrade along in the stationary column; see Figure 1 (p. 9). With each step, each soldier in each moving column encounters one comrade in the stationary column but *two* comrades in the oppositely moving column. Now imagine that each soldier represents an indivisible minimum unit of length and that each step represents an indivisible minimum unit of time. Surely we can ask the question: at what instant and in what position did the two moving columns align so that each soldier was alongside the next (rather than the next-but-one) soldier in the adjacent moving column? If we can sub-divide the time for the step and the space between steps there is no problem at all. For they will meet after half a step. But we have supposed that there is no such thing as half of one of our units of length or time – since they are

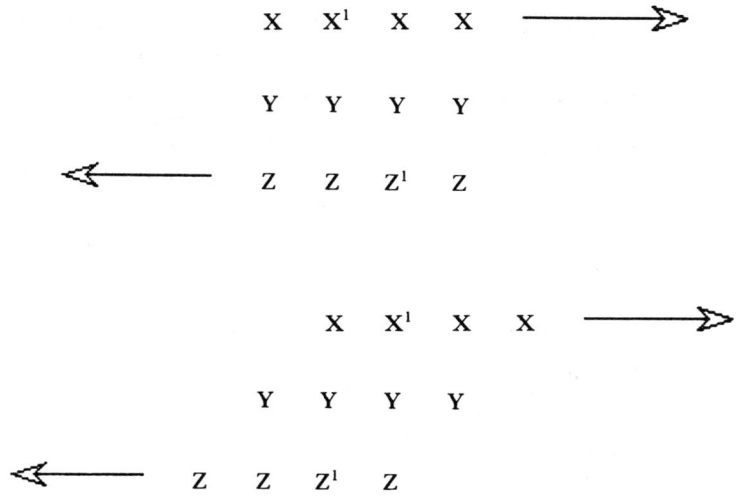

Moving rows paradox. Two rows (X and Z) move by a stationary row (Y) as shown. In the top diagram, X^1 and Z^1 are in adjacent columns, X^1 to the left and Z^1 to the right. An instant later, X^1 and Z^1 have shifted their positions, so that they are still in adjacent columns but with X^1 now to the right of Z^1 as shown in the lower diagram. Zeno's problem is this: when and where were X^1 and Z^1 in alignment vertically? Given that the change of position took place in the shortest possible time, we cannot say that they were in line in half this time. And, because the change of position involves the shortest possible distance, we cannot say that they were in line when they had moved through half this distance.

Figure 1 Zeno's moving rows or stadium paradox

indivisible minima. So either the question is unreasonable (and why should this be?) or we are wrong to suppose that space and time consist in indivisible minima.

In the first two paradoxes, Zeno tries to illustrate the absurdity of believing that a line may be divided up into progressively smaller chunks *ad infinitum*. And there is something seductive in his argument. For how can I move from A to B when I first must move to some point in between? And whatever point I choose and no matter how many times I do this, there is *always* going to be yet another point in between. Zeno warns us against saying that sooner or later I must reach the smallest possible 'indivisible' distance. For this discrete view of space too will generate problems, as demonstrated by the fourth paradox. Some writers approach Zeno's paradoxes with confidence, saying that just a little modern calculus will be sufficient to dispel any worries which the paradoxes may produce.[3] Ian Stewart identifies the central issue in Zeno as the way we think of infinitesimal quantities; and says that only in the last hundred and fifty years or so have we begun to see the problem in a way that helps us to resolve the paradoxes without too many qualms. Stewart asks:

> Can a line be thought of as a sequence of points? Can a plane be sliced up into parallel lines? The modern view is 'yes', the verdict of history an overwhelming 'no'; the main reason being that the interpretation of the question has changed.
> (Stewart 1987: 66)[4]

Mathematicians now seem to have few worries about continuous sub-divisions. What has changed is their attitude towards infinitesimal quantities. Such quantities are not regarded as extensionless points in space or in time. If we regard points as having no extension, then we fall victim to Zeno's fifth paradox: that of plurality – said by G.E.L. Owen and others to be Zeno's primary concern and to underlie the other four paradoxes.[5] Indeed Owen argues that we should regard the paradoxes as providing a coordinated attack on the reality of space, time, and motion. The first two paradoxes challenge the idea that space and time can be continuously sub-divided and the second two attack the notion that there are indivisible minima of space and time; so that Zeno's overall judgement may be summarised thus: 'no method of dividing anything into spatial or temporal parts can be described without absurdity.'[6] The fifth paradox discourages us from regarding the end result of some continuous sub-division as either an extensionless quantity like a point or a quantity with some definite if minute extension:

5 The paradox of plurality

Zeno, according to Simplicius, asks how even an infinite number of extensionless distances could add up to a finite distance and how an extended body can consist of an infinite number of parts (geometrical points?) which themselves have no extension; such a distance or such a body must be infinitely small – i.e. it must be just like its constituent parts: extensionless.[7] Yet if we allow these constituent parts to have some finite size – however small – then the body must be infinite in size.[8]

Owen points out that this paradox, taken together with the first four, may be seen as providing reasons for Zeno's view of the world as a single global entity rather than as made up of parts, whether these are indivisibly small or continuously divisible. As soon as we start to sub-divide we run into difficulties. So the sensible thing to do is to resist the temptation to divide the world up at all! Zeno's world is a single body which may not be sub-divided in any way without absurdity.

INFINITESIMALS AND LIMITS

Must we accept Zeno's conclusions? The answer seems to lie in our attitude towards the 'end' result of an unending process of sub-division, to the idea of infinitesimals. It is a mistake to regard them as having some 'constant' value whether this be the 'zero' of extensionless objects or points, or whether it is the non-zero value of the shortest possible distance or time. In both cases we would fall straight into one or other of Zeno's traps. We need a different approach if we are to avoid the traps altogether. The way out was first suggested by the French mathematician Cauchy in 1821: he introduced the idea of a limit; and the notion of the infinitesimal was absorbed into this more coherent concept.[9] And, some thirty years later, Weierstrass showed that we could move the debate from the realm of geometry to that of arithmetic, from ideas of spatial and temporal distances to those of functions. Instead of talking about ever-decreasing distances along a straight line, we could talk with a little more rigour about infinite series converging on limiting values in terms of functions and real numbers.

The problem may be highlighted by considering how we should answer this question: what speed does the athlete have at any given instant? If we think in terms of infinitesimals with a 'zero' value, then the equation for the speed of an object (distance ÷ time) collapses into nonsense – the speed of any moving object considered in this way will always be zero divided by zero! So, instead of saying that we may describe the motion of the athlete by reference to infinitesimal distances and times, we should calculate the speed of the athlete at any instant in terms of how the object is moving in the immediate neighbourhood, as shown by the

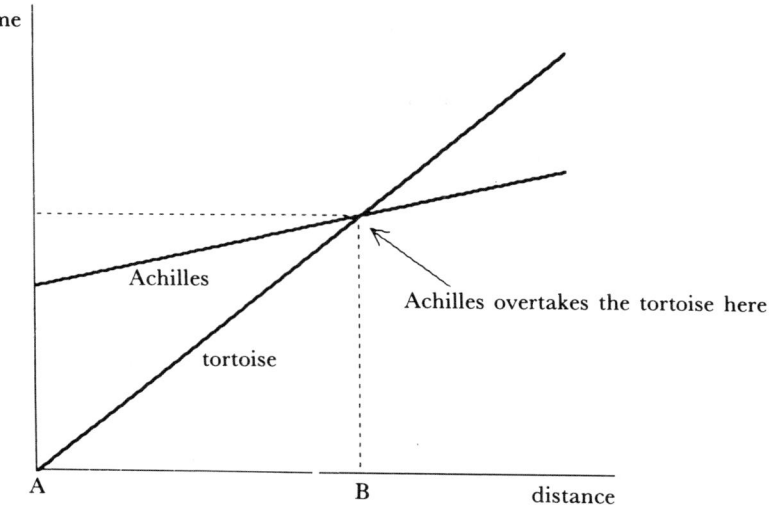

Although Achilles starts the race after the tortoise, because his speed is greater than that of the tortoise, he overtakes the tortoise at the point shown. The speed of Achilles (distance ÷ time) rather than the decreasing distance between the two is the key to the problem.

Figure 2 Distance-time graph comparing Achilles with tortoise: idea of velocity

mathematical function describing the athlete's motion. By considering smaller and smaller neighbourhoods, we typically reach a limiting value for the function – the 'instantaneous' speed. We get our answer by considering what happens as we *approach* the instant, not by asking what is happening *at* the instant. Similarly, we consider whether or not Achilles overtakes the tortoise and whether or not the athlete may run from A to B by thinking in terms of what happens as Achilles approaches the tortoise and as the athlete approaches the end of the racecourse; see Figure 2 (above).

So, using these ideas, we may give the following provisional responses to Zeno's worries about a continuum which may be continuously sub-divided:

1 The functions describing Achilles' and the tortoise's motions show that, when Achilles is in the immediate neighbourhood of the tortoise, Achilles' speed is greater than that of the tortoise and he therefore overtakes it.[10]

2 When Zeno's athlete attempts to run from A to B, the athlete

will indeed need to cover half the distance, then a further quarter, and so on; but the function describing the series of distances run by the athlete converges upon a natural limit: the total distance AB.

3 The idea of an instantaneous 'snapshot' of a moving object (e.g. an arrow) does not by itself carry with it any idea of motion; only when we consider its immediate spatio-temporal neighbourhood may we give any sense to the idea of a moving arrow – but once we do this the idea of motion is quite coherent (many action photographs show moving objects against a blurred background to convey the idea of speed – such pictures would capture the arrow not at an instant, but over a short period of its flight).

But we still face some important questions:

1 Are we entitled to say that a converging infinite series has a sum given the fact that an infinite sequence has a limiting value? For although we may agree that the *limiting value* of the sequence of partial sums:

(1/2), (1/2 + 1/4), (1/2 + 1/4 + 1/8), (1/2 + 1/4 + 1/8 + 1/16), . . .

is 1, we do not thereby have sufficient reason to say that the series:

1/2 + 1/4 + 1/8 + 1/16 + . . .

has 1 as its *sum*. We might accuse mathematicians of fudging the issue when they assure us that the limiting value of such a sequence is also the sum of a related series. So, although we may say that the sequence of distances from A run by the athlete has AB as its limiting value, this need not commit us to the view that the series may be summed at all! For such a summation seems to involve an infinite number of additions, and we might regard such an addition as at best implausible or at worst logically impossible.

2 To what extent may we say that the mathematical concepts employed above apply to the physical world? For, even if we allow that an infinite series may have a mathematical sum, this is no reason for us to agree that we may apply this procedure to the physical world with impunity. For example, we may use the idea of a simple arithmetical sum when adding one quantity of

money to another; but, when adding velocities in the Special Theory of Relativity, a different procedure is required. So we might ask: to what extent are mathematical and geometrical concepts and structures *strictly true* of the physical world? Why should abstractions apply literally to the physical world?

These questions are now addressed by Thomson, who challenges us to contemplate a super-task consisting of an infinite number of tasks.

THOMSON'S INFINITE SUPER-TASK

James Thomson asks us to imagine a lamp which may be switched on and off an infinite number of times in a finite time. If we set aside the question of whether or not it is physically possible for an infinite number of such tasks to be performed in a finite time, we may still ask with Thomson whether or not it is logically possible. We may easily imagine Thomson's reading-lamp with its switch in the off position and him switching it on then off then on and so on. If it is switched on at time zero and off after one minute and on again after another 30 seconds and off again after a further 15 seconds and so on, then we might think that after two minutes we would have completed an infinite number of switching operations. But Thomson asks:

> at the end of the two minutes, is the lamp on or off? ... It cannot be on, because I did not ever turn it on without at once turning it off. It cannot be off, because I did in the first place turn it on, and thereafter I never turned it off without at once turning it on.
>
> (Thomson 1954: 5)[11]

And this, Thomson tells us, is contradictory. He concludes that we could not in principle carry out such a super-task. Sainsbury says in *Paradoxes* that Thomson's conclusion is unwarranted.[12] He begins by distinguishing between those moments when the switching tasks are being performed (the T-series) and that first moment *after* the super-task has been completed (T*). Following Benacerraf's argument in 'Tasks, super-tasks and the modern eleactics',[13] Sainsbury then maintains that 'for any moment in the T-series, if the lamp is on at that time there is a later moment in the series at which the lamp is off; and vice versa'.

However, nothing follows from this about whether the lamp is on or off at T*, for T* does not belong to the T-series. The T-series is closed at one end (time zero) and open at the other 'end'. This means that although we can identify a first task at time zero, we cannot give any sense to the notion of a last task – the openness of the T-series guarantees the possibility of the tasks continuing an infinite number of times. The time T* does not occupy any point at this open end and is therefore independent of the T-series. But since T* is independent of the T-series, Sainsbury points out that our 'specification of the task speaks only to members of the T-series, and this has no consequences, let alone contradictory consequences, for how things are at T*, which lies outside the series' (Sainsbury 1988: 15).

So Sainsbury concludes that Thomson fails to demonstrate that the idea of a super-task is logically absurd. Two clear implications of Sainsbury's argument are:

1 that the lamp is either on or off at T*; and
2 that we cannot say either at the beginning of the super-task or once it is under way just how things will turn out at T*.

Hence, there is no way to predict the state of the lamp at T*, whatever its state at the outset.

THE PARALLEL TASK PARADOX

Suppose we ask an operator to carry out the super-task twice in succession in exactly the same way each time. Then there is no reason to suppose, on Sainsbury's view, that the lamp would be in the same state at each of the moments T* after the two tasks are completed. Otherwise we would always be able to predict the final state of the lamp. We may sharpen this problem as follows. Imagine now two lamps and one operator for each lamp. We ask both operators to attempt Thomson's super-task at the same time. Both lamps are off and at time zero both operators switch their lamps on. After one minute the operators switch the lamps off; after 30 seconds both lamps are switched on together; after 15 more seconds the lamps are off again; and so on. If we grant the point that T* lies outside the series, we may also grant that at T* each lamp will be either on or off. But are we forced to conclude that both lamps will be in the same state? Given that the operators begin together and continue together with the lamps flashing on and off in unison, we might expect them to finish together with the lamps in precisely the same state. However, what happens during the T-series is, we are told, independent of what is the case at T*. As Sainsbury claims, *nothing* concerning T* follows from our specification of the task to be carried out because this specification relates only to times within the T-series. The fact that the lamps are initially in the same state is irrelevant, since the moment at the start of the super-task lies within the T-series. And the fact that the operators continue together does not help, because our instructions to them relate solely to times within the T-series. So why should we expect the lamps to be in the same state at T*? We are left with the unsatisfactory conclusion that two people always in step during an infinite sequence of tasks may be out of step immediately after the sequence has 'ended'.

Such a conclusion seems to involve us in rather more than an *empirical* puzzle about the way things will turn out with such 'parallel' super-tasks. There seems to be a rational if not a logical inconsistency. If the super-tasks run in parallel and in step at all times during the series, then we have no reason at all to suppose that this pattern could be broken when the super-tasks are over. By accepting the distinction between the T-series and the (independent) time T*, then there is also now a reason to suppose that the pattern may be broken at T* when the tasks are over. The two lamps may be in different states at T* because there is no connection between the states of the lamps during the switching operations and their states at T*. Given this lack of connection, the chances of the lamps being in the identical states at T* seem to be the same as the chances of them being in different states at T*. And this appears to be an acceptable reason for believing that the pattern may be broken. Hence, acceptance of the independence of T* from the T-series leads us to a direct conflict with our apparently reasonable expectation that two 'parallel' operations should always remain in step not just up to the end of a task but also when the task is over.

The difficulty involved in this problem seems to derive from the fact that there is no limiting behaviour of the situation to which we might appeal in order to dissolve the puzzle. When Zeno's athlete attempts to run from A to B at a uniform speed – first passing through the half-way point, then the three-quarter-way point, and so on – the mathematical function which describes the athlete's progress can be represented on a graph of distance against time as a straight line. The natural limit of the line may be unambiguously defined as point B. Again, there is a natural limit for a lightly damped simple pendulum; in this case the graph of displacement from equilibrium against time shows that the amplitudes of the oscillations decrease towards the limiting value of zero displacement from equilibrium; see Figure 3 (pp. 18–19). But the lamp system has no such natural limit. For there is no preferred way for us to extend any mathematical function, which we employ to describe the behaviour of the system across the open end of the T-series to T^* itself.

It might be thought that this lack of limiting behaviour presents us with a resolution to any worry we might have concerning the parallel super-task. For it seems that we now have an explanation of the fact that two lamps which start out together can end up out of step. The mathematical function describing the system simply fails to determine which of two possible states each lamp will be in at time T^*. And why should this be so problematic? There are many, many situations in physics where we cannot uniquely determine outcomes.

However, the parallel super-task raises more difficulties than that of mere uncertainty as to the final outcome. Indeed, the source of any worry we might have is the conviction that both lamps should end up in the same state, whatever that might be. Of course, we have a degree of uncertainty as to how things will turn out. Yet the uncertainty is limited to that which we would have about one lamp – we just cannot say whether the final state will be on or off. Nevertheless, as Sainsbury reminds us, we do know that it will be in one of these states. But we *also* have compelling reasons for believing that the lamps will be in the same state.

Of course, it might be said that the peculiar circumstances of the task are such as to raise doubts about whether or not the lamps will be in the same final state. Perhaps quantum effects might break the symmetry. However, we need offer no reason other than considerations of symmetry between the two lamps to justify the belief that the lamps will be in the same state. But we do need some reason to suppose that the symmetry might be broken. If we were to provide even a sketch explanation for the broken symmetry, we would be supplying some kind of link between the lamps in operation during the infinite task and the lamps at T^*. But we have seen that the argument against Thomson proceeds on the assumption that there is no such link between the T-series and

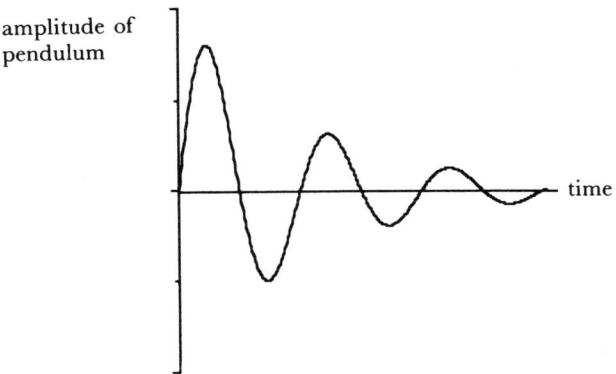

A lightly damped simple pendulum will swing to and fro, with its amplitude (maximum distance from the centre of the swing) decreasing. In theory, the amplitude will become closer and closer to zero but the pendulum will never actually stop. Hence the pendulum will move through a infinite distance – but only in an infinite time.

Figure 3(a) Graph of infinite distance in infinite time (pendulum)

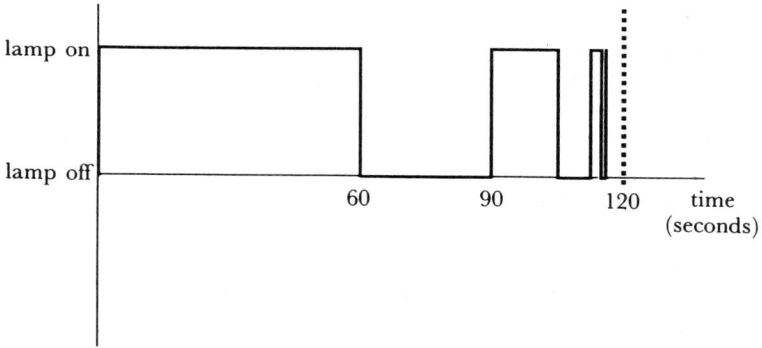

The lamp is switched on and off – each switching action happening after just half the time of the previous action. The 'infinite super-task' should be complete after 120 seconds. The finger pushing the switch on and off should therefore move through an infinite distance in a finite time – given the fact that the switch has to be pushed an infinite number of times.

Figure 3(b) Graph of infinite distance in finite time (Thomson's lamp)

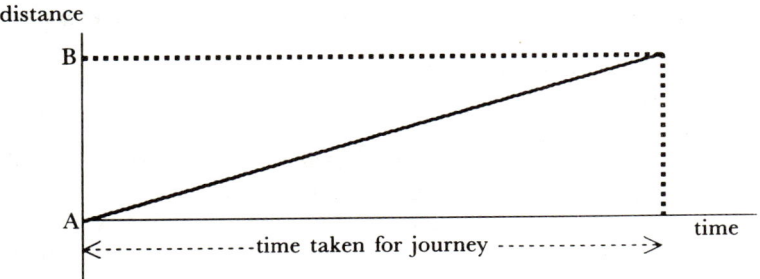

Zeno's athlete travels from A to B – a finite distance – in a finite time.

The natural limit for the amplitude of the simple pendulum is zero. The natural limit for distance run by the athlete is B. But there is no such natural limit for the switch on the lamp: it may only be 'on' or 'off'.

Figure 3(c) Graph of finite distance in finite time (Zeno's athlete)

T*, that this series and T* are independent. If this independence is challenged, then Thomson's original argument against such tasks must be faced.

If we wish to deny the genuinely paradoxical nature of super-tasks, then one of two options seems to be open to us. We may be forced to challenge our rational expectations concerning the outcomes of such tasks. Or we may maintain that only those paradoxes involving logical contradictions are genuine paradoxes. But there seems to be no reason at all to take the first option. And, together with Sainsbury, we have plenty of reasons to resist the second option – in his introduction, a paradox is described as 'an apparently unacceptable conclusion derived by apparently acceptable reasoning from apparently acceptable premises'.[14] And, despite its lack of any outright logical contradiction, the super-task certainly falls into this category. So we should follow Thomson's advice and rule out such tasks in principle. Although the mathematics of the continuum might seem to provide us with a powerful strategy to attack Thomson's original paradox, we are given insufficient resources to resolve all the ramifications of such super-tasks. Perhaps the moral of this tale is that, when we conjure up empirical fictions, we should not be too surprised when our stories end unhappily, even if they do have impeccable mathematical credentials.

ABSTRACTIONS AND THE PHYSICAL WORLD

What lessons does the lamp paradox hold for our view of Zeno's paradoxes? Unlike the lamp operators, athletes like Achilles seem to have only one task ahead: to run from A to B. Hence, we cannot say that the lamp paradox provides any direct support for Zeno's attacks on the continuum: the task facing Achilles does not seem to be a super-task. Yet someone might still want to say that there is a *sense* in which Achilles is complying with instructions similar to those given to the lamp operators: first go to the half-way point, then a further quarter, and so on. But are we really asking Achilles to carry out an infinite number of tasks in any sense? This just does not seem to be so. Max Black draws our attention to the central issue here:

> Achilles is not called upon to do the logically impossible; the illusion that he must do so is created by our failure to hold separate the finite number of real things that the runner has to accomplish and the infinite series of numbers by which we describe what he actually does. We create the illusion of the infinite tasks by the kind of mathematics that we use to describe space, time, and motion.
>
> (Black 1950: 101)[15]

The link between the lamp paradox and Zeno's paradoxes lies, not in the nature of the tasks under discussion, but rather in the shared concerns about the status of the continuum.

Benacerraf and Sainsbury follow an approach to Zeno's worries about the continuum which focuses our attention on the consistency of the mathematics used to discuss the problem. The mathematics allows us to think consistently in terms of limits. But, as soon as we try to apply the mathematics to the physical world, the problems begin – as we saw in the case of the parallel super-task. As Salmon points out 'the applicability of mathematical concepts to specific types of physical phenomena is not an automatic consequence of their consistency'.[16] We imagine Achilles proceeding in regular strides down the track. No one asks the athlete to move through an infinite number of ever-shortening steps. Any notion that the athlete moves through an infinite number of distances derives from the *mathematical* possibility of subdividing the total distance AB as many times as we wish (and more *ad infinitum*). So any claim that the athlete is running through an infinite number of physical distances relies on the questionable assumption that what seems to be true of an abstract mathematical domain should also apply strictly to the physical world. In the mathematical domain, we may readily imagine a series closed at one end and open at the other or a length divisible continuously *ad infinitum*; but in the physical world lengths are measured by rulers and other such instruments – and we always read off a discrete finite amount, a rational number for any distance we measure.

Could I not make smaller and smaller measurements at least *in principle* if not in practice? The key phrase here is 'in principle' – using it runs together mathematical and physical possibilities. What I can do in practice is constrained by what is physically possible. Although I cannot always decide whether or not a given task or situation is physically possible and I can also acknowledge that advances in science and technology may open up more possibilities, I generally do know when something is possible and when something is not. There may be no hard-and-fast distinction. But distinction there certainly is. Of course we may play mathematical games here, and imagine what it would be like if . . .; but, in the physical world, it is practice rather than principle which dictates possibilities. The problem with taking too vigorous an empirical line here is obvious: we do use mathematics in our physical theories with a tremendous degree of success. But even an enthusiastic acknowledgement of the utility of mathematics should not carry with it the presumption that because it works it must *strictly* apply to the physical world.

Both mathematical abstraction and physical observation point us towards the same answers to Zeno: Achilles overtakes the tortoise, the athlete reaches the goal, and the arrow is in flight. Mathematics provides us with the idea of a limiting value in an attempt to capture what is happening physically. Here, as elsewhere, mathematics and geometry provide us with the materials for models of the physical world. But models are abstract analogies and not the world itself. The spatio-temporal continuum is one such model. And, as with every analogy, there are points of difference as well as points of similarity.[17] Just as the lamp supertasks should make us worry about the application of mathematics to the physical world, the lesson for us provided by Zeno's paradoxes is that we should not be too quick to stress the similarities between the continuum model and the observed world at the expense of what may be crucial differences. But we should also beware the temptation to substitute an indivisible space and time model for the continuum. Even if the mathematics and geometry used in such a model is consistent and powerful enough to escape Zeno's fourth paradox – the attack on 'atomism' – the question of strict applicability still remains.

An interesting argument, discussed by W.H. Newton-Smith, highlights this issue and should provide a further reason for us to avoid treating the continuum with too much respect.[18] The argument recalls the essential feature of the Duhem–Quine thesis: that 'physical theories can be at odds with each other and yet compatible with all physical data even in the broadest sense' – i.e. that all possible empirical evidence may not be sufficient to commit us to any one theoretical view.[19] We have already noted that the measurements which we make are always rational numbers – numbers which may be formed by the ratio of integers as fractions. No matter what distance we measure along, say, a metre rule, that distance will be a definite fraction of the entire length. But an irrational number such as $\sqrt{2}$ cannot be represented by any fraction. And no possible measurement could deliver an irrational number. Yet the structure of the continuum used in Newtonian (and relativistic) physics assumes that the distance between any two points in space or time may be represented by an ordering of real numbers – rational *and* irrational.

Newton-Smith suggests an alternative theory to Newton's mechanics (he calls this Notwen's theory) which could be based on an ordering of the rational numbers alone. Hence, Notwen's theory would eschew the 'smooth' continuum structure involving both rational and irrational numbers for space and time. Of course, Notwen's theory could not use the differential equations of the calculus to analyse motion since these equations assume a smooth continuum. However, it could use difference equations – rather more cumbersome, but no implicit reference to irrational numbers would be needed. Notwen's theory would certainly not be elegant nor would it be easy to use. But it would be consistent with all possible empirical evidence! For every measurement we make is in terms of rationals. The data given in observation and experiment are not enough to determine our choice of theory despite the unattractiveness of Notwen's theory. The complexities of such a theory, apparent even when we consider one-dimensional orderings, would increase dramatically in the context of two-, let alone three- or four-dimensional structures – indeed, it is far from clear how we would carry the basic idea of a rational ordering from one to the familiar context of three spatial dimensions.

So we might still choose to use Newtonian theory. However, the selection would be made on pragmatic grounds, and not because the mathematical and geometrical concepts employed indicate something about the 'true' nature of the physical world. But, even if we commit ourselves to the idea of the continuum as the basis of our physical theories of space, time, and motion, we must still decide on the appropriate 'global' geometry. For, as we shall see in Chapter 3, we no longer have a straightforward 'choice' between Euclid and nothing. And, once again, we shall face the problem that the empirical evidence is not always sufficient to force any particular theoretical view of the world upon us.

2
CLOCKS, GEOMETRY AND RELATIVITY

INTRODUCTION

In June 1905, Einstein published a paper entitled 'On the electrodynamics of moving bodies' in which he reflects upon difficulties arising from Maxwell's electrodynamics and the Newtonian ideas of space, time, and motion. In this short paper the essential ideas of the Special Theory of Relativity (STR) are set down. These ideas, Einstein tells us, have a 'peculiar consequence':

> If at the points A and B of [a given inertial frame of reference[1]] K there are located clocks at rest which, observed in a system at rest, are synchronized, and if the clock at A is transported to B along the connecting line with velocity v, then upon arrival of this clock at B the two clocks will no longer be synchronized; instead, the clock that has been transported from A to B will lag . . . behind the clock that has been at B from the outset.
>
> (Einstein 1905: 153)[2]

Einstein's prediction here allows us to say this: if someone were to leave the Earth in a fast-moving spaceship, then on that person's return, say ten years from now in Earth terms, any clock carried in the ship would have ticked off fewer seconds than any clock on Earth. And, since a person is, in an important sense, no more than a biological clock, the space traveller too would have aged less than those who stayed behind. STR also allows us to say that the faster the traveller moves in the round-trip, the greater the age difference between the traveller and those who stayed behind.

Some people have regarded such predictions as 'paradoxical', finding it at least counterintuitive if not downright contradictory that people inhabiting the same universe should not all age at the same rate. But this 'paradox' of STR seems to run deeper than any simple conflict with our everyday intuitions. For STR predicts that, just as I would say that your clock runs slow when you are in relative motion with respect to me, you too would say the same of my clock, that it runs slow and your own clock is ticking away as usual – so which of us is right? Any straightforward assertion that 'moving clocks run slow' is clearly going to be problematic, for you and I will be in disagreement about which clock is moving and which clock is running slow. Since all inertial frames have the same status in STR, we cannot say: 'my frame of reference is a more appropriate standard for clocks than your frame, so my clock will always behave correctly, and your clock, so long as it is moving with respect to me, will be running slow'. For STR blocks my giving any frame such preferential status. Yet, despite the symmetry of our respective situations, frames and descriptions, STR also predicts that a clock returning from a 'round-trip' will indeed have ticked off fewer seconds than the clock that stays behind – that in such a situation only one of us will be right in saying that the other clock is ticking more slowly than our own clock. And this apparent clash of predictions does indeed seem to be puzzling.

In April 1922, the Collège de France debated the implications of STR. And Einstein's predictions about clocks in the paper of June 1905 caused the participants great concern. The French philosopher, Henri Bergson, responding to the Collège debate, said that, if we apply the predictions of STR to the real world, then we shall indeed 'end in absurdities of paradoxes'.[3] He argues that when a space traveller or a clock moves away from us at high speed the moving objects are no more than phantoms of the physicist's imagination. The mathematical calculations of STR apply only to such phantom images. When the real traveller and clocks return, we shall see no difference between them and the people and clocks left on Earth. But Bergson reveals his prejudices about time when he asserts that there may be a 'multiplicity of imaginary times' in STR but there is nevertheless 'a single, real time'. Bergson wants Einstein's relativistic cake, and yet he is determined to eat it baked à la Newton. The predictions of STR offend his Newtonian belief that there is a single global time frame in which all the different kinds of clocks (including the human body) tick at the same rate. So STR is accepted by Bergson as mathematically interesting, but physically inappropriate.

Even if we were to feel some sympathy for Bergson's prejudices, we need to be aware that STR's predictions concern far more than objects in high-speed motion. I can make the same prediction for any object moving relative to me, even at the slowest of speeds. My chances of actually noticing the effect are very low except when objects are moving at speeds close to that of light. The easiest way to confirm STR's predictions about clocks is to examine high-speed objects such as mu-mesons travelling close to the speed of light. Mu-mesons are unstable particles with a reasonably definite lifetime in the laboratory, but travelling at some 0.99 of the speed of light they last about nine times longer.[4] But we can also measure the effect and confirm the prediction using clocks travelling on ordinary passenger jets, as shown by Halfele and Keating in their experiment of October 1971 when clocks joined the jet-setters: the time elapsed on airborne clocks in a round-the-world trip were compared with the time elapsed on clocks located in the US Naval laboratory – and the measured time differences agreed closely with the predictions of STR.[5]

So Bergson is wrong when he argues that those returning after a journey will be exactly the same age as us and this applies just as much to the objects of everyday experience as it does to the high-speed life of mu-mesons. We shall see that there is some truth in his belief that, when we are observing objects in motion, we are observing 'phantoms' in some sense. However, only a more detailed acquaintance with the language of STR will help us to realise in what sense Bergson may be right about this. And this more detailed study will also help us to see in what sense if any this prediction about clocks is paradoxical.

A brief, preliminary survey of the ideas of motion in Newtonian space and time and STR spacetime is provided in Figure 4 (pp. 27–33). The next section will then develop these ideas within the context of STR. Some readers may wish to move on immediately to p. 33.

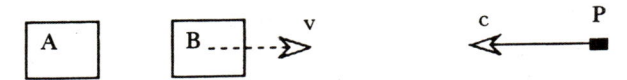

A spacecraft A is at rest relative to an observer O. Spacecraft B is moving away from A and O at a steady speed v. An object P is moving towards A at a steady speed c.

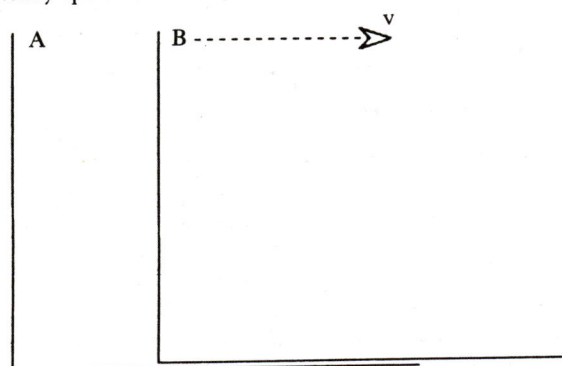

The diagram here illustrates two frames of reference: that in which A is at rest, and that in which B is at rest. Because B is moving at a steady speed v towards P as shown, the frame of reference in which B is at rest is also moving at v relative to the frame of reference in which A is at rest.

The people in spacecraft A say that P is moving towards them at a steady speed c. But what speed do they say will be measured by those in spacecraft B? The calculation is easy in Newtonian physics: we simply take into account the relative velocity between the two spacecraft v, so that those in B are said to measure a speed c − v for P.

Thus Newtonian mechanics gives us a straightforward way of comparing events and measurements in two or more frames of reference. But Newtonian physics places no restriction on the value obtained for the speed of light by different observers.

When we also specify that different observers moving uniformly must all measure the same speed for light, we run into problems. Imagine that P is a pulse of light travelling towards both A and B. A says that it is moving at a steady speed c. We may no longer say that those on B will measure a different speed c − v! For they too, given the restriction, must also measure c.

This means that we can no longer rely on Newtonian mechanics when we compare the measurements made by different observers − so long as the restriction is justified, as it seems to be by all the available empirical evidence. We need a new way of comparing events and measurements in different frames of reference − and the Lorentz transformations of the Special Theory of Relativity provide us with just that.

Figure 4(a) From Newton to STR

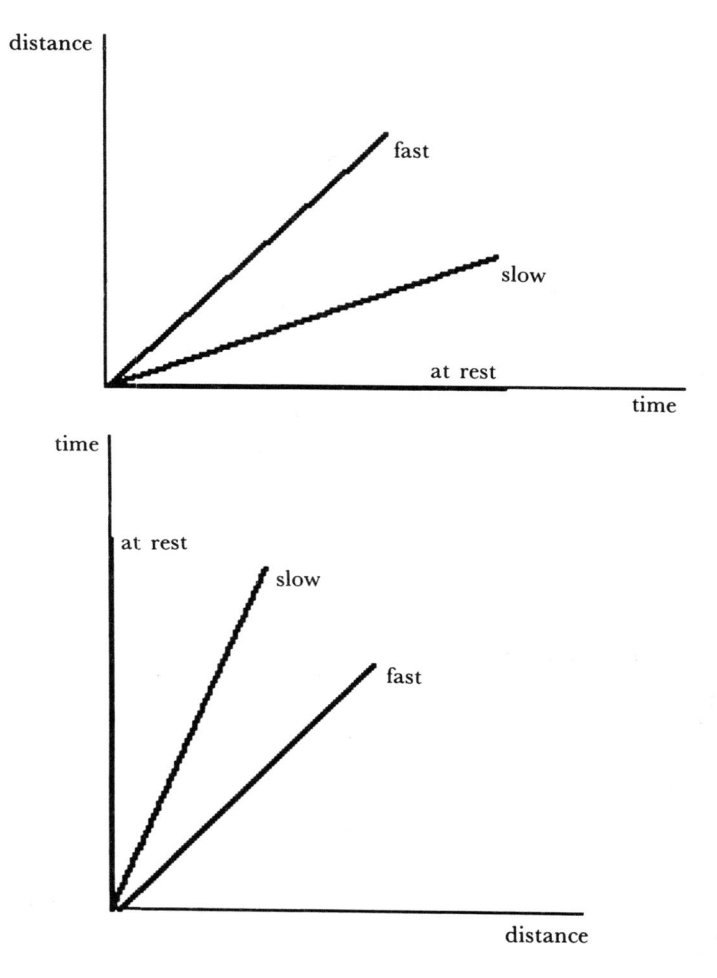

In a distance–time graph (top), the steeper the slope of the line representing a moving object or signal, the faster that object or signal is moving. Stationary objects are represented by a horizontal line along the time axis. In a time–distance graph (bottom), the time axis is now vertical and a stationary object is represented by a vertical line along this axis. The faster the object or signal, the further it moves away from a stationary object in a given time.

Figure 4(b) Time–distance graphs

Spacetime diagrams are usually drawn as time–distance graphs, with the time axis vertical and with just one spatial dimension (left/right) represented by the horizontal axis.

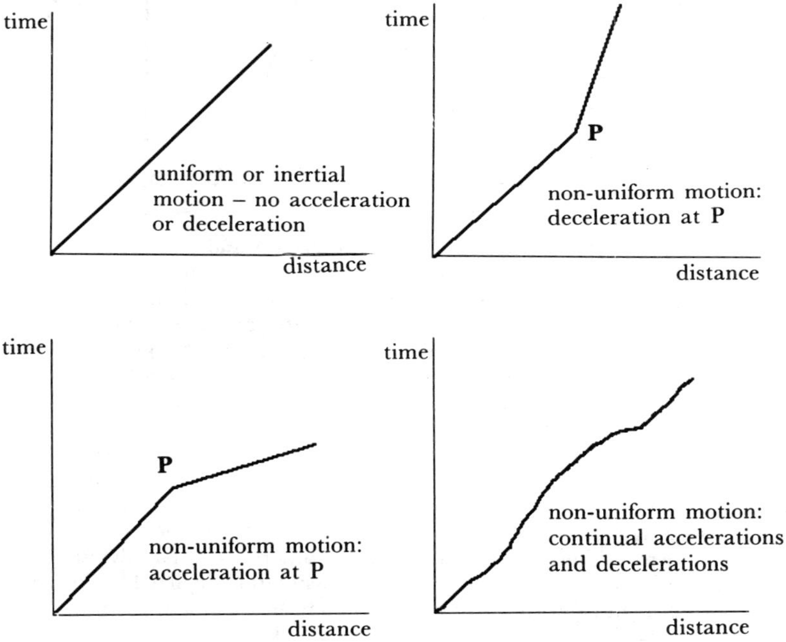
time
distance
uniform or inertial motion – no acceleration or deceleration
time
distance
P
non-uniform motion: deceleration at P
time
distance
P
non-uniform motion: acceleration at P
time
distance
non-uniform motion: continual accelerations and decelerations

In uniform (straight-line) motion, there is no acceleration or deceleration. Any object or signal in uniform motion moves at a constant velocity relative to any other object or signal in uniform motion. If the constant relative velocity between two objects is zero, then those two objects are at rest with respect to each other. They may, however, still be moving relative to other objects in uniform motion.

Each line representing an object or signal in a spacetime diagram is called the world line of that object or signal. Each point on a world line identifies a unique spacetime location for the object or signal represented by the line.

Figure 4(c) Ideas of velocity and acceleration

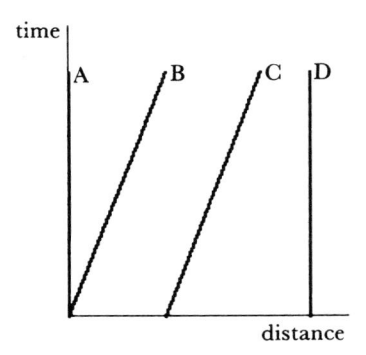

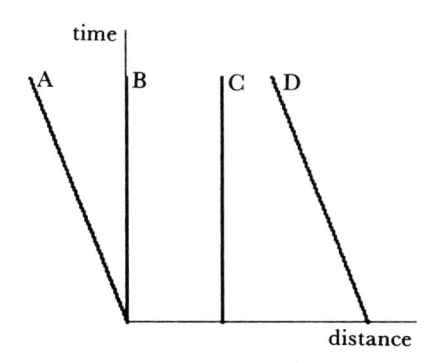

The diagram on the left illustrates two spacecraft A and D at rest with respect to each other. This is shown by the fact that the distance between their world lines never changes. But two other spacecraft B and C, moving away from A and towards D, are also at rest with respect to each other – here too the distance between their world lines does not change. When world lines are parallel – as in the case of A and D and also of B and C, we say that the objects or signals represented by the world lines are in the same frame of reference.

The spacetime diagram itself fixes a particular frame of reference – that in which all objects with world lines parallel to the vertical time axis are at rest. But we may redraw any spacetime diagram by taking any object or signal moving uniformly as the main reference – with its world line parallel to the time axis. The diagram on the right shows B and C parallel to the time axis. Notice, however, that the physical situation depicted in this new diagram is identical with the original situation. B and C are still moving steadily away from A and towards D. And A and D are still at rest with respect to each other – shown by the fact that their world lines are still parallel. It does not matter which object we focus upon and take to be stationary with respect to the time axis – the physical relationships between all the objects remain the same.

A reference frame defined by a uniformly moving object such as A or C is called an inertial frame of reference. The object represented by B occupies first one inertial frame and then after its change of direction another inertial frame of reference.

When an object changes from one inertial frame of reference to another it experiences inertial forces. Hence any deviation from straightness for a world line indicates that the object represented by the world line has changed frame and has experienced inertial forces as it accelerates or decelerates in changing frame.

Figure 4(d) Reference frames

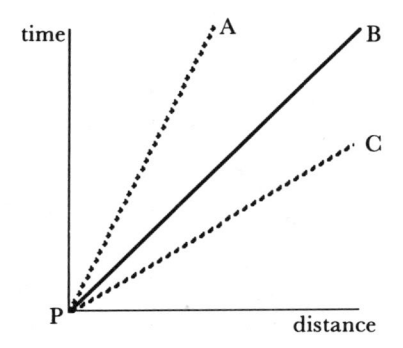

If B is the world line of a light signal, then A is the world line of a signal or an object travelling at less than the speed of light. But, if C represents any moving object or signal, then the object or signal must be moving faster than light.

P is a point in spacetime: imagine that P represents the explosion of a star. When the star explodes, the light from the explosion will travel in all directions away from the point of explosion.

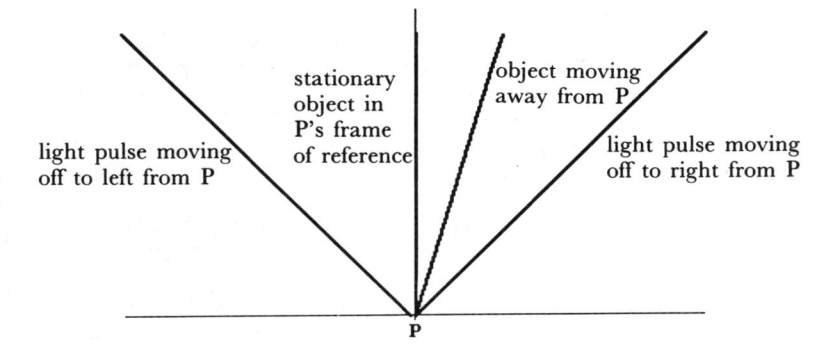

This spacetime diagram allows us to capture all the information about the explosion of the star in one spatial dimension. From this point onwards, we shall draw the world lines of all light pulses at 45 degrees to the time and distance axes as here. If the scale is one centimetre to one second on the time axis, the scale on the distance axis will be one centimetre to roughly 300,000,000 metres on the distance axis – since the speed of light is just under 300,000,000 metres per second.

Figure 4(e) Basic ideas of spacetime diagram

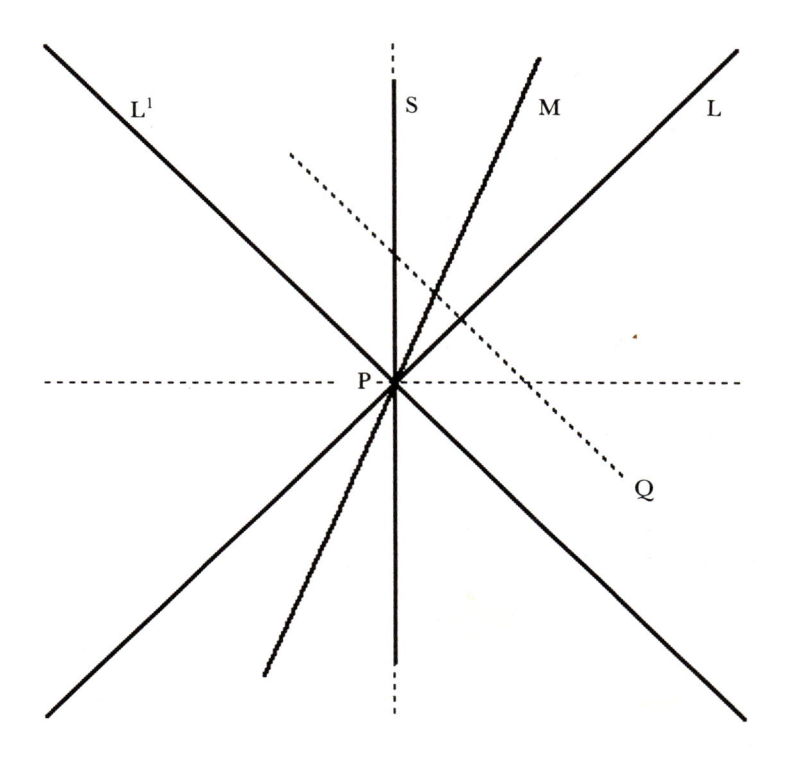

A spacetime diagram drawn from the point of view of S: i.e. drawn in S's frame of reference

S represents a stationary object. The point P represents an event such as a moving object M passing by S. M moves closer to S, meets S at P, and then moves away from S as shown.

L and L^1 represent two pulses of light – L moving in towards S from the left, and L^1 from the right. They meet S at point P and then move away as shown.

We have yet to detect any object or signal which can travel faster than light. So as far as S is concerned, a signal from point Q just cannot reach S until after P. For this reason, S regards the past of P as being defined by the portion of spacetime within L and L^1 as these two pulses converge upon S at P. Similarly, S regards the future of P as all those events which might be reached by a signal or object moving away from P, i.e. that portion of spacetime within L and L^1 as these two diverge from S at P.

Figure 4(f) Spacetime diagram showing past and future of an event P

CLOCKS, GEOMETRY AND RELATIVITY

MY TIME AND YOUR TIME

The Lorentz transformations (LT) give us the mathematical tools which we need to analyse motion in the spacetime of STR. Here, without going into the mathematical details of the LT, I shall try to spell out their substance and rationale. If a spaceship passes me at a steady (inertial) speed, then the LT allow me to predict what would be observed from any other inertial perspective: for example, from another spaceship either travelling alongside the first or rushing by at twice the speed of the first. In this way, I may 'view' the passage of the first spaceship from any inertial frame I choose. So the LT allow me to do far more than stay on Earth and simply measure the time passing spaceships appear to take to travel across the solar system. I may also predict how long the journey would seem to be if I were travelling inside the spaceship itself. Of course, this would hardly be interesting if every view were to be identical and every time prediction were to correspond to 'Earth time'. But STR breaks free from the global time perspective of Newtonian physics with everything in the universe running together at the same rate. With STR, the perspective from which we view an event or sequence of events becomes of critical importance when we make predictions about how a clock will tick compared with some other clock.[6]

If there is one time of which we can be sure, then it is our own time – the time we use to measure events in our own locality. No one is going to have much success in persuading us that we cannot rely on clocks for such local measurements. If we cannot depend on them, then it would be difficult to accept any predictions at all about how clocks are behaving in distant places. Fortunately, it is this 'personal' time-keeping which provides the touchstone for all time predictions in STR. When we use the LT we are trying to discover just how some distant and possibly moving clock (as far as we are concerned) is behaving in its own locality *when compared with our own*. And we are trying to avoid the claim that how any distant clock *appears* to be ticking is the way it *really is* ticking. A telescope might help us look and say how a clock *seems* to be running, but only the LT can compare the behaviour of the clock in its locality with a clock in our vicinity in a neutral way. With the LT we may deliver the same judgement upon all distant and/or moving perspectives as we deliver upon our own perspective. The LT transform away local prejudices and in doing so help us to see what is happening at some specific distant location when compared with our own or any other location (and vice versa).

If I stand in Parliament Square looking up at Big Ben and measure the time between the clock reaching twelve and one, then my own clock should show that one hour has passed.[7] One hour for Big Ben is one hour for me too: let us call this time measured by me T_0. But what about someone flying by in a jet – will one hour for Big Ben be one hour for them as well, when Big Ben will seem to be moving with respect to them? According to STR, any clock moving at a speed v relative to me will measure a time T between two events in my frame; and this time is given by the LT which lead to the following equation:

$$T = \beta T_0$$

where the factor β depends upon the relative speed v.[8] If my stationary clock measures the time T_0 (= one hour), then, using the above equation, I may calculate the time T measured by the moving clock. We may note that β has the value 1 when v = 0, i.e. when the object is at rest with respect to us. For low speeds relative to us, therefore, β is approximately 1 and the times T_0 and T are very nearly the same. This of course is why we see little disagreement about most time measurements here on Earth. But, if the clock is moving at all, then in every case the time T_0 measured by the stationary clock will be less than the time T measured by the moving clock. So one hour for the stationary clock is not one hour for the moving clock. As you fly by in a jet, your clock will read T between the two events: (E_1) Big Ben reaching twelve and (E_2) Big Ben reaching one; and, if T_0 is one hour, T will be *greater* than one hour. As far as you in your jet are concerned, Big Ben will be running slow. And it will be running slow because it is moving with respect to you, i.e. it is slow relative to your clock.

Now imagine a rather large jet carrying Big Ben, yourself, and a clock. I am stranded on the ground with my clock. We now consider exactly the same two events as before: E_1 – Big Ben at twelve; and E_2 – Big Ben at one. Because you and your clock are now in the same inertial frame as Big Ben, i.e. because there is no relative motion between you, your clock, and Big Ben, one hour for Big Ben will now be one hour for your clock. But we can only find the time my clock measures between E_1 and E_2 by using the LT. And the same equation as before tells us that I now measure a time greater than one hour with my clock! Big Ben takes one hour from your point of view, but it takes more than one hour from mine –

Big Ben will seem to be running slow. Again, it will be running slow because it is in motion, this time from my point of view, i.e. it is running slow relative to my clock.

Big Ben acts as a standard clock against which we may synchronise our own clocks. We may summarise the Big Ben story in terms of a standard clock. If a standard clock is stationary with respect to a clock to be synchronised, then one hour for the standard clock is one hour for the clock to be synchronised. But, if the standard clock is moving from the point of view of the clock to be synchronised, then the LT show that the standard clock and all clocks in the same frame as this clock are running slow from the perspective of the clock to be synchronised. So, only clocks in the same frame may synchronise perfectly.[9] And *any* clock (electronic, clockwork, or biological) in a frame moving relative to me or any other observer will run slower than my clock. It is the symmetry of two possible descriptions of two clocks in relative motion that has disturbed some people: two observers in relative motion will each say exactly the same about each other's clock: that it is running slow relative to their own clock.

How can the two descriptions *both* be correct? So long as we remain wedded to the idea of a single 'real time', we are unlikely to find the results of the LT credible. For this idea carries with it the assumption that there is a global time standard against which all clocks may be judged. But STR challenges precisely this assumption. Because there is no global standard to which we can appeal, no inertial perspective may be singled out as 'the correct' view. Hence, neither of the two descriptions above may be deemed 'correct' – rather, they are complementary descriptions of motions within a spacetime characterised by STR and the LT. In STR, distances in spacetime are invariant: such distances are called spacetime intervals; they are distances from one location in spacetime to another.[10] Although we will never disagree about the spacetime interval, we shall in general disagree about the length of an object or the time between two events unless our perspectives are the same. This notion of an invariant interval leads us to the idea of 'proper time' – or local time as it is sometimes called. When a signal travelling at less than the speed of light is observed to move between two points in spacetime, the interval between those two points is said to be timelike. If the signal between the two points in spacetime is travelling at the speed of light, then the interval is said to be lightlike or null; and, if a signal travelling faster than light would be needed to connect two points in spacetime, then the interval between the points is said to be spacelike; see Figure 5 (p. 37).

THE PARADOX OF THE TWINS: FOR EVER YOUNG?

The story of the paradox of the twins may now be told. Identical twins take part in an experiment to test STR's predictions. One twin remains on Earth and the other takes a round-trip in an extremely fast-moving spaceship to some distant star; when the trip is over and the twins meet to compare notes, the one that stayed at home is older than the traveller, just as all the clocks on Earth have ticked off more seconds than a clock taken with the travelling twin. The faster and further the journey, the more marked will be the differences between the traveller and the Earth-based twin. But each of their notes include the claim that the *other's* time is stretched out or dilated. If we leave the story at that, then we might think that there is a degree of paradox here. On the one hand, there is a remarkable symmetry here: the LT allow each twin to make the same claim – it is the *other* twin who is ageing slowly. But, on the other hand, the two situations do seem to be asymmetric: one of the twins is quite definitely older than the other – older in the sense that that twin has lived through more seconds. So only one twin does age more slowly. Yet this seems to conflict with the view already expressed that there is no 'correct' view. Surely the twin on Earth is right when that twin says that the other twin is ageing more slowly; and the traveller is simply wrong to claim that the twin on Earth is ageing more slowly! We now have two main choices available:

1 we might question the assertion that the traveller would have aged differently from the twin who stayed behind; or
2 we might look for some asymmetry between the two situations which will allow us to explain the difference in ageing.

But the empirical evidence mentioned in the opening to this chapter suggests that the assertion made is correct: clocks taken on round-trips do seem to tick off fewer seconds, and, as we have said, people may be considered as biological clocks. So the evidence suggests the two would have aged differently. The second choice seems to offer more promise. For there does seem to be an asymmetry which has been glossed over by our failure to tell the

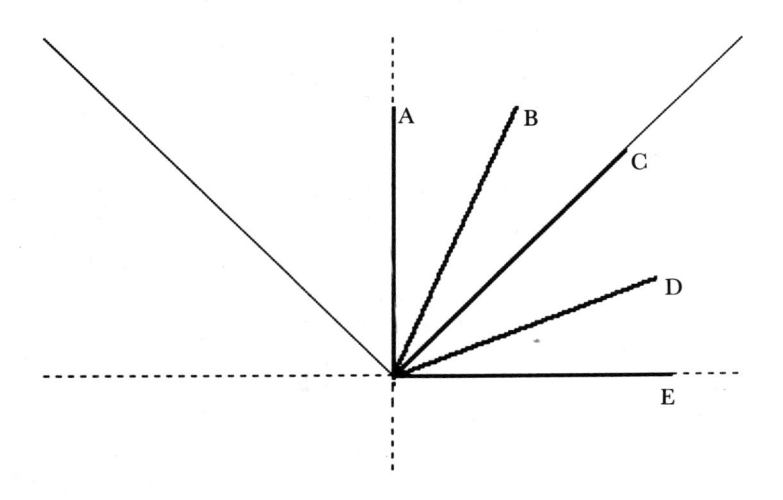

A: this world line represents an observer taken to be at rest – this observer judges B to E to be in motion; the line shown is a timelike interval.
B: this world line represents an object taken (by A) to be moving at a speed less than that of light; the line shown is a timelike interval.
C: this world line represents a light signal; the line shown is a null interval.
D: this world line represents a signal or an object moving faster than light (according to A); the line shown is a spacelike interval; it is generally supposed that no signal or object could move between points connected by a spacelike interval – whether or not this is so is an empirical problem and is not ruled out a priori by STR.
E: this world line represents a signal or object moving 'instantaneously' in A's judgement; this line too is a spacelike interval.

Figure 5 Null, timelike, and spacelike intervals on a spacetime diagram

whole story about the twins. Yes, it is true that each twin, using the LT, will make the same claims about each other. Their situations may seem to be symmetrical, for we might think that, just as the twin on Earth may say that the Earth is at rest and that it is clearly the spaceship which moves first away from and then back to the Earth, the travelling twin may also consider his or her spaceship to be at rest, and view the Earth as steadily moving first away from and then back to the spaceship. But, unlike the twin on Earth, the travelling twin will not be able to consider himself or herself at rest for the entire journey!

When the spaceship turns around, the frame of reference of the travelling twin will change. If, in order to make a round-trip here on Earth, someone attempts to jump from the back of a truck travelling at 30 m.p.h. in one direction to the back of a truck travelling at the same speed but in the opposite direction, then it will soon be obvious to the person trying to do this that they will change frame by doing so. For the two trucks are in frames of reference travelling at 60 m.p.h. with respect to each other. And when objects change frame, forces act upon them – the greater the change, the greater the acceleration or deceleration involved, and the greater the resulting force. Anyone who has jumped from a bus moving at just 10 m.p.h. relative to the pavement might be able to guess just how foolhardy sudden, dramatic changes of frame can be. The travelling twin will experience such a change when the spaceship turns around – as it slows down and then speeds up for the return leg of the journey. Given that the effects of a round-trip will only be clearly visible as far as the twins are concerned if the traveller's journey is at speeds close to that of light, the change of frame will indeed be dramatic – from a tremendously high speed away from the Earth to an equally high speed towards the Earth – especially if we restrict the time available for the change of direction. But we shall suppose that the traveller survives the change.[11] The spacetime diagrams in Figure 6 (p. 39) illustrate the two situations: only one of the twins may be represented by a world line which never bends or changes direction. For a world line which is both straight and unidirectional signifies that the object represented by the world line is moving inertially and its frame of reference is unchanging. A world line which changes direction indicates that the object represented by the world line experiences a change of frame, and therefore it also experiences an acceleration and inertial forces. The travelling twin's world line changes direction but the world line of the twin on Earth does not. And the diagrams show that this is the case whichever twin we take to be at rest at the start of the journey. So, we may break the apparent symmetry in the story of the twins by pointing out that only one twin changes frame, only one twin decelerates and then accelerates, and only one twin experiences the inertial forces due to the change of frame.[12]

What is responsible for this difference in ageing? The straightforward answer is the fact that the spacetime paths of the twins are different. In STR, we may no longer treat motions as taking place

The twins rejoin each other at this point

time

P

P

A B

A B

The twins begin to move apart here

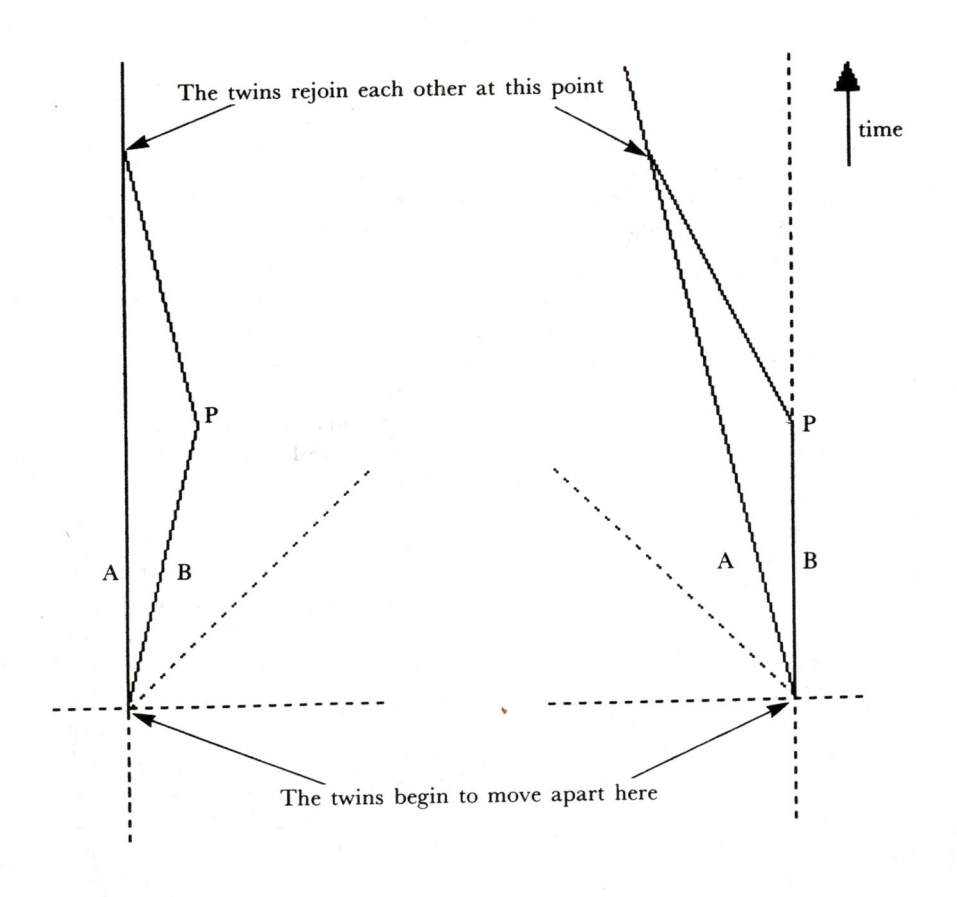

The spacetime diagram on the left is drawn from the point of view of the twin A who stays at home. This twin experiences no inertial forces – the world line representing this twin remains straight. But the travelling twin B experiences inertial forces at the turnaround point P. The diagram on the right illustrates the situation from the travelling twin's point of view. Although this twin might be able to say that it is the other twin who is moving as the distance between the two twins increases, only one of them experiences an acceleration and only one of them experiences inertial forces – and this happens to the travelling twin at the turnaround point. The change of frame experienced by this travelling twin demonstrates that there is only a superficial symmetry between the two perspectives.

Figure 6 Twins paradox spacetime diagram

in a vast spatial arena with some independent global cosmic clock ticking away in the background. A full description of any motion will always specify how an object moves in spacetime, taken as a composite entity. The world line of an object provides such a description. And because the world lines of the twins are not the same, there is no reason for us to expect clocks which were synchronised at the start of the journey, when the respective world lines intersect, to synchronise at the end of the journey, when the world lines next meet. When we embrace the spacetime perspective of STR, why should we regard this as any more problematic than the fact that two people who take different spatial paths (e.g. the high road and the low road) between two fixed points on a map will in general travel different distances, each taking a different number of paces to complete their respective journeys?

Yet the 'paradox' does seem to leave us with a puzzle. If two identical clocks synchronise at the start of an experiment, but disagree at the end, then we might be inclined to look for a physical reason for the disagreement. But the answer given above seems to explain the differences in terms of spacetime geometry rather than in terms of matter and forces. So it is tempting to say that the difference in world lines, and therefore the difference in ageing, is explained by the fact the traveller experiences inertial forces when the propulsion system operates to accelerate and decelerate the spaceship. These forces are the physical cause of the difference. This need to find a material explanation of the effect seems to derive from a positivist outlook. Typically positivists argue that only those things which we can observe in some straightforward way should be included in our physical descriptions of the universe. We can 'see' neither space nor time nor spacetime geometry. So positivist inclinations would naturally lead us to discount explanations of physical effects in terms of geometry and to look for something more concrete such as inertial forces with their origin in some kind of material interaction.[13] Unfortunately, STR does not completely satisfy such a positivistic desire. For we may readily reconstruct the tale of the twins in such a way as to leave out all mention of inertial forces!

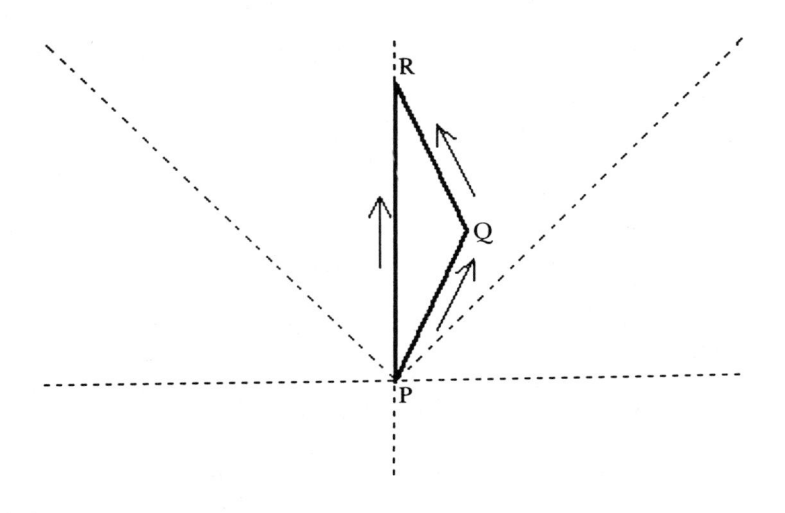

The diagram shows two ways to get from P to R in spacetime: via the spacetime path PR (along the vertical axis) and via the path PQR.

Figure 7 Different spacetime paths for twins

FROM TWINS TO TRIPLETS

The spacetime diagram in Figure 7 above shows the spacetime paths of the twins. The world lines diverge from point P and meet again at R. The change of frame occurs at point Q. So there are two distinct spacetime paths: PR; and PQ ... QR. Instead of one person following the second path in its entirety, now imagine that one traveller 'moves along' the line PQ such that at no time does that person experience inertial forces; and that another traveller 'moves along' the line QR again without experiencing inertial forces.[14] As before, we imagine that the person on Earth, represented by PR, also experiences no inertial forces. So now there are three people involved in our story — triplets instead of twins, each carrying an identical clock.[15] The three clocks involved may be labelled:

C_1 – for the clock on Earth;
C_2 – for the clock with the traveller represented by PQ; and
C_3 – for the clock with the traveller represented by QR.

We may now retell the tale in terms of a second thought experiment as follows:

1 At the intersection of the lines PR and PQ, the clocks C_1 and C_2 synchronise, and, as far as we can see, the two triplets involved appear to be the same age.
2 At the intersection of the lines PQ and QR, the clocks C_2 and C_3 synchronise, and, again, the two triplets involved here seem to be the same age.
3 At the intersection of the lines PR and QR, the clocks C_3 and C_1 do not synchronise, nor do the ages of the two triplets involved at this point agree. The clock C_3 lags behind C_1, and the travelling triplet represented by the line QR is younger than the triplet on the Earth line PR.[16]
4 At each intersection P and R of the world lines, the relative speed of those represented by the intersecting world lines is the same so that, from the point of view of the triplet on Earth, the velocity of the triplet represented by PQ is equal in size but opposite in direction to that of the triplet represented by QR.[17]

The lack of synchronisation at R is explained by the difference in spacetime paths; see Figure 8 (p. 43). However, we may no longer look to inertial forces to explain the asymmetry. For no one involved in this thought experiment experiences such forces.

Some writers have argued that the paradox of the twins cannot be satisfactorily resolved within the context of STR, and that only by using the General Theory of Relativity (GTR) may we properly explain the asymmetrical effects; see Bohm (1965).[18] Whatever reasons we might have for moving to a theory like GTR in order to give an account of the two thought experiments, it is wrong to suggest that we should do so because STR cannot deal with accelerated motion. This claim about STR is false, as French, amongst others, points out:

> Because Einstein developed a whole new theory ... based upon the dynamical equivalence of an accelerated laboratory and a laboratory in a gravitational field, it is sometimes stated or implied that special relativity is not competent to deal with accelerated motions. This is a misconception. We can meaningfully discuss a displacement and all its time derivatives [e.g. acceleration] within the context of the Lorentz Transformations.
>
> (French 1968: 153)[19]

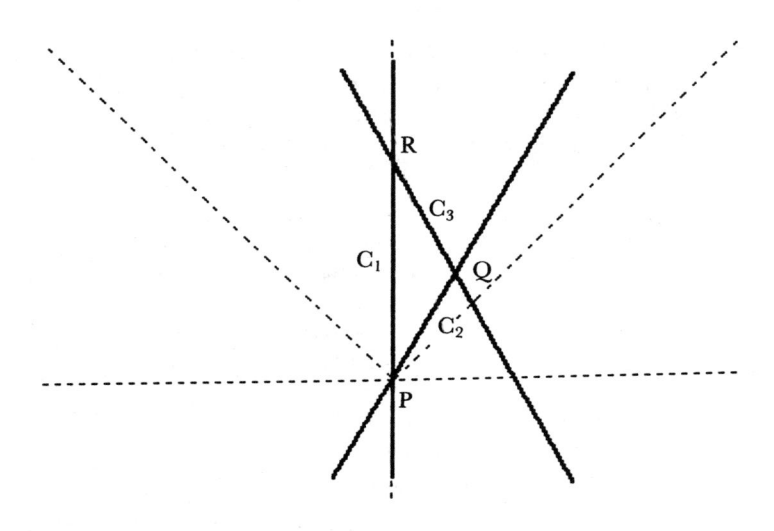

The clocks C_1 and C_2 synchronise at P; C_2 synchronises with C_3 at Q; but, at R, C_3 lags behind C_1.

Figure 8 Triplets paradox spacetime diagram

What STR does not do is to give an account of motion in gravitational fields, i.e. in arbitrarily curved spacetimes. For curved spacetime, GTR is indeed required; the essential details of curved spacetime will be explored in Chapters 4 and 7.[20] There is, interestingly, a gravitational time dilation effect predicted by GTR: from the point of view of someone in low gravity, a clock in a high gravity area will run slow. Given the equivalence between gravitational and inertial effects which is part of the foundations for GTR, it might be thought that this is reason enough for us to shift from STR to GTR. Yet there is no firm evidence that gravitation is an issue either directly or indirectly in the 'paradoxes' above. And, even if we are tempted to move to GTR to explain the asymmetry in the story of the twins because inertial forces are involved in that thought experiment, the tale of the triplets indicates that the asymmetry has nothing to do with forces – for no forces are involved in any way.

This produces a dilemma for anyone with a positivistic bias. The positivist is generally eager to find some concrete observational origin for any physical effect. This, as we shall see in Chapters 6 and 7, is how the problem of the source of inertial forces is handled

– the positivist claims that inertial forces arise when objects accelerate relative to some average of the material contents of the universe. But, since we can imagine this 'triplet' thought experiment taking place in an otherwise empty spacetime, what might we cite as being responsible for the asymmetry? Each of the three paths involved is inertial. And we have no reason to single out any one path for preferential treatment in any way. Even if we do give preferred status to one path, we would clearly have some non-material reason for doing so. The positivist might still try to argue that only in the context of a complete dynamical theory which sets all motions within a realistic, gravitational context may we talk sensibly about such effects. However, even if we set the thought experiment within the observed universe, since none of the participants experiences any forces and all three may be moving relatively to the material contents of the universe, we may only point to the fact of relative motion to explain the resulting asymmetry. And why, from a materialist point of view, should objects in relative uniform motion not behave in the same way? What is true materially for one object may also be true for the other objects involved. The only significant differences seem to be the *directions* of the three motions. So there seems to be little encouragement here for the positivist, who might perhaps regard the thought experiment as genuinely paradoxical – or give up his or her positivistic bias.[21]

PHANTOMS OF PERSPECTIVE

Bergson argued that the travelling twin would be no more than a phantom in the physicist's imagination and that on his or her return to Earth ages and clocks would all agree. We can now see why STR does not support such a view. I may not say that a person in motion relative to me is a phantom, just because we may both make the same claim about each other. All we may say is that neither claim has any priority over the other, for neither claim is 'the correct' one. And there is no reason why clocks travelling between the same points but along different spacetime paths should agree. We appeal to geometry to resolve the 'paradox' – to the geometrical ideas of spacetime paths and world lines which present the motion of objects in the only way possible in STR: within the union of space and time.

Throughout the tales of the twins and the triplets, I tried to avoid talk of what each participant actually sees. This would complicate the stories tremendously. For, in a certain sense, the people involved in such experiments would not see each other 'as they really are'. When we use telescopes to view distant events we do not see things 'as they really are' – and it is not at all clear what might be meant by this phrase. There are three problems which arise from any simple-minded reliance on telescopes:

1. Different telescopes in different locations and moving at different speeds will not, in general, agree on the 'facts' about some sequence of events.
2. The 'facts' which they reveal are always from the perspective of the viewer and not from that of the viewed; for example, because time elapses between the emission of any signal from a source and its capture by the telescope, the event viewed always lies to the past of the act of observation – we see things 'as they were'.
3. Other effects may prevent the telescope from giving an accurate observational description of what is happening at the distant locality – for example, the Doppler effect may cause us to attribute properties to a distant event which it does not in fact possess.

Nevertheless, we may make corrections for all such problems, so that the results obtained from the LT agree with observation. Yet these problems may still tempt us into accepting that distant events and objects are phantoms in a sense – for they do not have the immediate tangibility of events and objects in our own locality, and we never see them except from our local and partial viewpoint. But, as indicated earlier in this chapter, this is exactly why the LT are needed.

The LT enable us to explore without distinction all events and motions within the spacetime of STR. They provide a mechanism by which we can overcome our locally bound perspectives. They help us to give to distant objects and events the same concrete status as those in our own locality. Bergson's phantoms are as solid and real as Bergson (was) himself.

3
TRAVELLING LIGHT

INTRODUCTION

Many attempts to measure the speed of light clearly involve the measurement of light travelling to a given point and back; so the speed measured is that for the 'round-trip'. For example, Fizeau's method, a standard experiment in the mid-nineteenth century, involves light travelling from a source to a distant mirror and back to an observer close to the original source; again, Foucault's method, which replaced Fizeau's approach as the standard procedure, also measures the speed of light travelling around a closed path.[1] Numerous actual and thought experiments have been suggested in order to help us measure the one-way speed of light. Wesley Salmon argues that all such attempts involve a fatal flaw: any experiment designed to measure the one-way speed actually involves either a round-trip measurement or some other problematic assumption.[2] Hence, we are told, the claim that the speed of light (*in vacuo*) in a given direction is always c is essentially a non-trivial *convention*. This has an immediate consequence for our definition of simultaneity. The Special Theory of Relativity (STR) demands a change in the way we think about simultaneity. The idea of a fixed 'invariant' speed of light is a key element in STR's account of simultaneity. The light cone, which represents light spreading out in all directions from an event, provides the foundation for all judgements of simultaneity.[3] We can no longer rely on some vast cosmic time-slice providing a plane of simultaneity for all observers. We use light rays to synchronise events and the judgement that two events are simultaneous depends on how light rays propagate between them. If we want to be sure that a nearby sequence of events synchronises with a

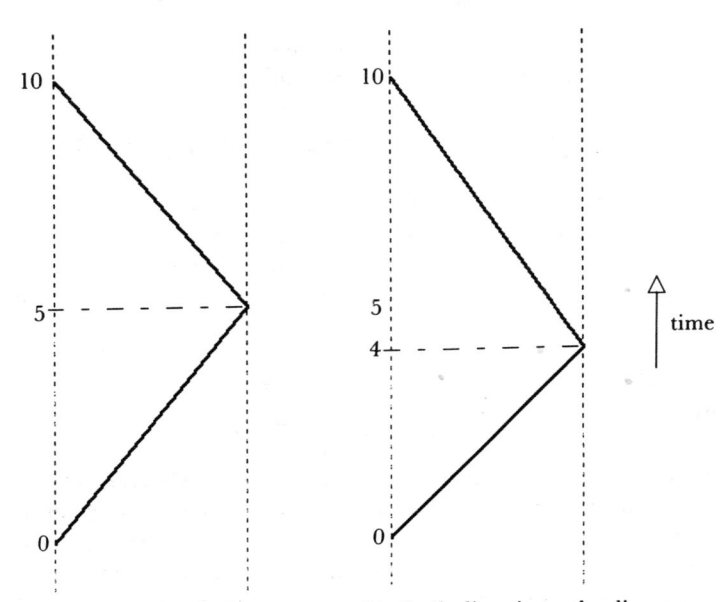

If the light signal travels at the same speed in both directions, the diagram on the left applies; but, if not, then the situation on the right might apply.

Figure 9 Planes of simultaneity in STR using light beams for synchrony

distant sequence in our frame of reference, then we typically bounce light rays back and forth to check that the sequences match. The shuttling light rays transmit the latest information about each of the sequences. If the round-trip takes 10 seconds, then we would generally feel confident in the assertion that the latest information received always relates to an event 5 seconds old. But we need to be sure that the time the light ray takes to travel from us to the distant sequence is the *same* as the time for the return trip. Otherwise our judgement of simultaneity between pairs of events in the sequences will be impaired; see Figure 9 above. Therefore, if our assertion that the light travels at the same speed in both directions is conventional, then our judgements of synchrony and simultaneity will also have conventional characteristics.[4]

Much that we say depends upon some convention or other. Before publication of a novel, its title is agreed. On publication, that title is accepted as the proper way of referring to the novel. But the book could quite easily have had some other title. The particular title given is just a matter of conventional agreement even though some titles may be more appropriate than others. Again, I accept and follow the general convention that the object before me is a Macintosh computer. But all such objects could quite easily have been called something else. What we call the novel or the machine makes no difference to the content of the book or the operation of the computer. Such conventions concern only the use of one name rather than another. However, some conventions might be far less trivial than these. In the first chapter, we observed that it is hard to justify the choice between continuous and merely dense spatio-temporal structures on the basis of empirical evidence alone. But, if we nevertheless decide to agree that space and time form a continuum, then our agreement on this point seems to imply something about the way the world is. In this case, such conventional agreement is far from trivial.

In arguing for the conventional character of at least some important scientific beliefs, some philosophers adopt an anti-realist 'conventionalist' viewpoint. Many philosophers and scientists are reluctant to admit that any major decisions about the nature of the physical world are based on conventions, agreed amongst the scientific community, rather than on independent empirical facts. They prefer to commit themselves to a realist perspective in which:

1 what science says about the world is at least approximately true – even at the theoretical level;
2 the truth or falsity of all scientific claims may be determined by independent reference to the way the world is and not by any pragmatic or conventional beliefs held by the scientific community; and, as a matter of historical fact,
3 the history of science reveals overall progress – a convergence on truth, with better and better approximations to the truth.[5]

Conventionalists may be 'realists' to a *limited* extent; they may believe that the world exists independently of our minds. They may accept that our theories are in some sense about the world and that what we say about observations (if not theory) may be true of the world. However, they are likely to remind us that we are sometimes faced with alternative theories and hypotheses which seem to be equally supported by empirical evidence. So they typically maintain that our choice amongst the alternative theoretical descriptions is essentially conventional. Their position seems to be motivated by a sceptical attitude towards our knowledge of the world: some realists would like to think that everything we say about the world may be determined as true or false, at least in principle; but many conventionalists argue that the scope of our knowledge is restricted in principle by the need to make conventional decisions. Some writers go further than this. They argue that all our theoretical choices have a conventional character: the empirical evidence can *never* tie us down to just one view; that is, our choice of theory must always be *underdetermined* by the empirical data.[6]

If empirical evidence turns out to be insufficient to settle a dispute about the physical world, then the realist position would be challenged. In this chapter and the next, we shall explore a number of situations which might be said to have a conventional character. We shall then be in a position to assess the strengths, and weaknesses, of the conventionalist challenge to realism.

Another issue connected with the speed of light concerns the possibility of signals or particles travelling faster than light. We shall therefore take the opportunity in this chapter to explore the problems associated with particles travelling faster than light. We shall find that relativity does not legislate against such particles, but that they do lead to some very strange physical consequences, particularly the possibility of travel backwards in time.

MEASURING THE SPEED OF LIGHT

Is Salmon right when he says that the one-way speed of light is a convention rather than an empirical fact? He reviews a wide variety of attempts to measure the one-way speed; but, at this stage, we shall discuss just one of the methods which he examines.[7] Despite initial appearances, this method assumes implicitly that the speed of light is the same irrespective of its direction.[8] In the early nineteenth century, the English scientist Thomas Young investigated the wave nature of light by observing the effects of light passing from a single source through two narrow slits in his celebrated 'two-slit experiment'. The overall effect may be displayed on a screen; and this shows firm evidence of interference between the light waves spreading out from the two slits. A regular series of light and dark bands or fringes, the results of constructive and destructive interference, is clearly seen. The distance between the bands depends upon the wavelength of the light used. Once we know this distance, a straightforward geometrical argument allows us to calculate the wavelength. Given that the wavelength of light is related to the speed of light, we may readily calculate this speed on the basis of the results of Young's two-slit experiment. On the face of it, this method appears to offer a straightforward measurement of the one-way speed of light: speed is given by the expression 'distance travelled ÷ time taken'. The distance is a straight line in one direction from slits to screen in Young's experiment. No round-trip distance *appears* to be employed in the calculation.

However, Salmon and others who support his views argue that in order to arrive at the figure for the one-way speed of light, we must assume that light travels at the same speed independently of the direction in which it travels! The relationship between the separation of the interference bands and the wavelength of the incident light in the two-slit experiment depends upon an important assumption about the propagation of light. The geometrical argument used by Young assumes that light spreads out *in the same way in all directions* from each of the two slits. Hence, if we use this method to measure the speed of light, we have already tacitly assumed that the speed of light is independent of its direction. We have not provided an objective way of determining the one-way speed; we have simply reconfirmed our convention that direction does not matter.

Those debating this issue of the alleged conventionality of the one-way speed of light ask us to focus on a fundamental issue which seems to lie at the heart of the problem: the structure of spacetime in STR. The standard method for synchronising events in STR's spacetime involves the transmission, reflection, and reception of light signals between two observers at rest with respect to each other: this method is called 'standard signal synchrony'. Imagine two such observers A and B: A sends a signal to B who holds a mirror which reflects the signal back to A without delay. A sends the signal at $t = 0$ and receives the reflected signal at $t = 2$. At what time according to A did B reflect the signal? Clearly, with a finite transmission speed for light, A does not see B at the moment of reflection as such: A must wait for the reflected signal to arrive. But it might seem reasonable for A to make the assumption that the time for the signal to travel out is exactly equal to the time it takes to return. This, of course, relies on the further assumption that light travels at the same speed in both directions: same speed, same distance, so same time! Hence, A may say that the moment of

TRAVELLING LIGHT

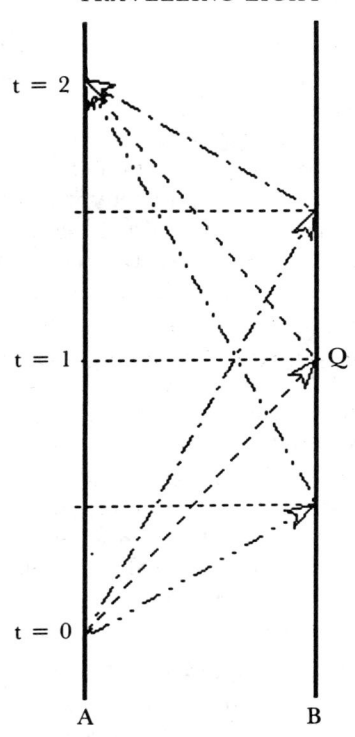

Any point on A's world line from $t = 0$ to $t = 2$ could be chosen as being simultaneous with the arrival of the light signal on B's world line.

We say that $t = 1$ is metrically simultaneous with Q – the arrival event. But the entire set of points from $t = 0$ to $t = 2$ is said to be topologically simultaneous with the arrival event – for Q might be located at some other position on B's world line; two other possible positions are shown in the diagram.

The choice of $t = 1$ as the appropriate metrically simultaneous event is a conventional decision according to Grunbaum and others.

Figure 10 Metrical and topological simultaneity

reflection is at $t = 1$ on A's clock. The plane '$t = 1$' then defines the plane of simultaneity for the 'reflection' event Q; see the horizontal dotted line from A to Q on B in Figure 10 above. Using signals in this way then allows all distant observers in the same frame to agree on which events are and which are not simultaneous with any other given event.

However, if we challenge the assumption that the speed of light is independent of direction, then A may not say, without some other assumption about the speed of light, just what time on A's clock is simultaneous with Q. Nor may we use standard signal synchrony to provide a general definition of distant simultaneity in STR. For no unique plane of simultaneity may be specified for any event whatsoever. Reichenbach provides a useful way of highlighting the problem here.[9] In general, the relationship between the times for the transmission (t_0), reflection (t_1), and reception (t_2), according to A, may be specified as follows:

$$t_1 = t_0 + \varepsilon (t_2 - t_0), \text{ where } 0 < \varepsilon < 1$$

Reichenbach points out that the case $\varepsilon = 1/2$ corresponds to that in which the speed of light is the same in both directions, i.e. the time t_1, supposed to be simultaneous with Q, occurs at precisely $1/2$ the time taken for the round trip ($t_2 - t_0$). So, when $t_0 = 0$, we may suppose that $t_1 = 1/2\ t_2$. Hence, when we assume the speed of light is independent of direction, we also assume that that $\varepsilon = 1/2$. Adolf Grunbaum clarifies the dilemma for observers such as A along the following lines:

1 There is an infinite number of events between $t = 0$ and $t = 2$ on A's world line which might be said to be simultaneous with Q;
2 we are so uncertain about which event is simultaneous with Q given the fact that no signal travelling faster than light may be used to narrow the choice;
3 therefore the set of points between $t = 0$ and $t = 2$ on A's world line may only be connected to Q by spacelike intervals — for example, the horizontal dotted line from A to Q on B is a spacelike interval representing the path of a notional 'instantaneous' signal.
4 this set as a whole may be said to be *topologically simultaneous* with Q;
5 a decision to stipulate the speed of light in any given direction will pick out a unique event from this set as being *metrically simultaneous* with Q; however,
6 any such decision must be conventional: for example, we may choose $\varepsilon = 1/2$ and appeal to the simplicity involved in our assumption that the speed of light is not dependent on direction, and then fix on $t = 1$ as being simultaneous; but, equally,
7 we may adopt a different convention, such that ε is not $1/2$, and

select some other point from the topologically simultaneous set; in general, any value may be selected for ε between 0 and 1;
8 hence metrical simultaneity depends upon a conventional decision which is forced upon us as a matter of fact given that the round trip speed of light is finite; see Figure 10 (p. 51).[10]

The immediate problem which this analysis suggests is this: can we develop a version of STR consistent with the belief that ε may take any value between 0 and 1? John Winnie seems to provide a convincing and consistent account of a theory which looks like STR except in its assumption of ε = 1/2 and the repercussions of this assumption.[11] The key difference between standard STR and this new version is that in the former we speak of the invariance of the velocity of light *in vacuo* in any direction as an essential (if conventional) idea in the theory; whereas in the latter we may only refer in general terms to the invariance of the round-trip velocity. However, David Malament argues that the conformal structure of spacetime in STR forces us to accept the relationship of simultaneity by the standard signal method. There are several levels of structure within any spacetime: metric, affine, conformal, and topological; see Figure 11 (p. 54) for a review of the properties associated with each level. The conformal structure underwrites the notion of an angle in spacetime and therefore provides us with the ability to construct planes of simultaneity at right angles in all directions to any world line. Conformal structure is regarded for this reason as more fundamental than the metrical features associated with measurements between points. Malament shows that the implications of dropping the assumption that ε = 1/2 are far-reaching, since we would have to revise our ideas about the conformal structure of STR's spacetime. Yes, it is true that we could in principle adopt some non-standard view. But the necessary revisions to the framework of our spacetime theories which such a move would require would be so great as to take us clearly outside of the context of STR.[12]

ABSOLUTE SIMULTANEITY?

Grunbaum supposes that the conventionality of metrical simultaneity follows from a matter of fact about the finiteness of light's round-trip speed. This supposition reveals an interesting assumption, which he shares with Wesley Salmon. Both he and Salmon

Metrical structure of spacetime
The metric of a spacetime gives sense to the idea of the distance between points and the length of a line. Once we have specified a metric in a given spacetime, we are able to talk sensibly about, not just distances and lengths, but straight lines and the angles between two straight lines. However, the notion of a straight line does not depend upon the ideas of distance and length, and so we may introduce two (more general) kinds of spacetime structure to capture the ideas of straightness and of angles.
Affine structure of spacetime
The affine structure of spacetime picks out all the straight lines in that spacetime. Since we define straight line motion as force-free motion, the affine structure allows us to distinguish the class of inertial motions from that of non-inertial motion.
Conformal structure of spacetime
Conformal structure gives sense to the idea of an angle between two intersecting straight lines.
Topological structure of spacetime
The topology of a spacetime captures the properties of the points of spacetime themselves: whether or not the points form a continuum, how many dimensions the spacetime has, whether or not the spacetime has boundaries setting its limits or holes within it, whether the time direction at any point is well defined, and so on. Hence, the topology of a spacetime is said to describe the general characteristics of the set of points forming the spacetime 'manifold'. This manifold is sometimes considered as the fundamental background arena into which the other three kinds of structure are introduced.

Figure 11 Synopsis of properties of spacetime structure: topological, affine, conformal, and metric

seem to believe that, if there were some signal which could travel to some distant point and back instantaneously, then the issue of the conventionality of simultaneity would dissolve. Like Salmon he accepts that:

> According to Newtonian mechanics . . . material particles can be accelerated to arbitrarily high velocities, well beyond the speed of light. If Newtonian mechanics were true, such particles could, in principle, be used to synchronize clocks and to establish relations of absolute simultaneity.[13]
> (Salmon 1980: 119)

And, like Salmon, he also acknowledges that 'We are not making any assumption about the relative one-way velocities of the fastest super-light signal that is employed.'[14] Presumably Salmon and Grunbaum think that such a signal would travel infinitely fast in one direction and infinitely fast in the other. But this need not be the case! For all we would know in such a case, according to Grunbaum's own position, is that the *round-trip* takes place instantaneously. If we allow the possibility of travel backwards in time, then the following situation seems possible:

1 at $t = 0$, the outward light signal travels at a finite speed forwards in time; but, after reflection,
2 the returning signal travels backwards in time and arrives at the starting-point, so that no time has elapsed at that point.

Alternatively, the outward signal might travel backwards in time, and the returning signal forwards to arrive once again at $t = 0$. Of course, it would be difficult to single out 'instantaneous' light signals for such special treatment; how light travels depends upon the metric of the spacetime and upon more fundamental features such as conformal structure. We would therefore expect all moving objects to behave in a similar manner.

So we would need to choose between a 'simple' Newtonian world in which 'time travel' does not take place and a more complex situation in which it does occur. Clearly, such a strange situation would require radical revisions not just in the account we give of the various structures of Newtonian space and time but also in our other physical theories. However, a number of possibilities for 'time travel' are raised by the General Theory of Relativity and other physical theories, and so perhaps we should not be too hasty in dismissing the idea as absurd despite the advice of some philosophers. But allowing such a possibility might also allow us to say that the direction travelled by any signal in time might be dependent on the spatial direction of the signal! Figure 12 (p. 56) illustrates three possible situations:

1 An infinitely fast signal travels from A to B and back along route 1, i.e. from M to Q and back; if the signal travels in all directions in the same way, then we can define a plane of 'transmission' orthogonal to A, i.e. at right angles to A in all directions, which is coincident with the plane of simultaneity $t = 0$; the events Y, M, and Q lie on this plane, which is common to every observer.
2 A signal travels from A to B and back along route 2, i.e. from M to R and back, such that the outward signal travels forwards and the return signal backwards in time; if we are able to construct a 'plane of transmission' consistently in such a spacetime, then in this case the plane would be angled 'downwards' to intercept C

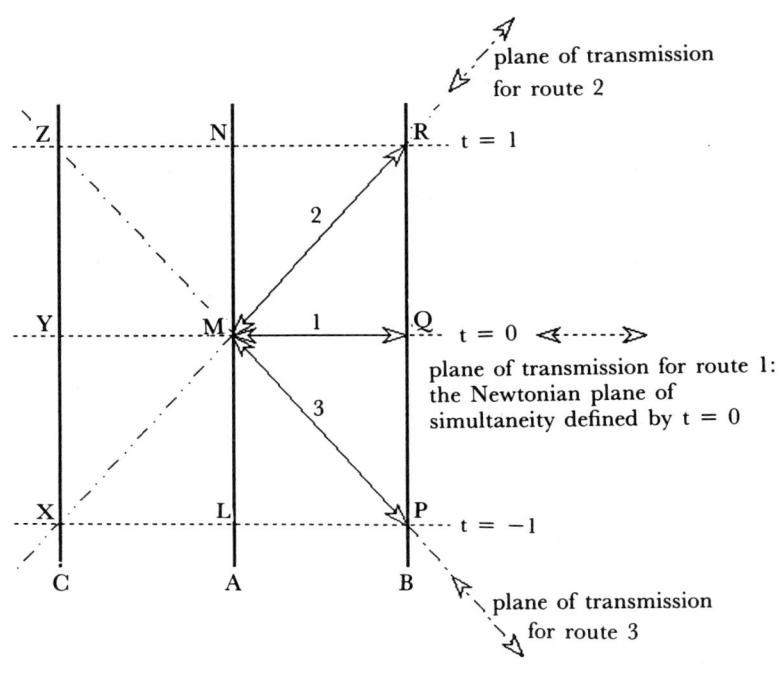

Figure 12 Absolute simultaneity

at X as shown; the Newtonian plane of simultaneity through the reflection event R (defined by the paths that one-way instantaneous signals would take) is $t = 1$, and the events Z, N, and R lie on this plane.

3 The signal travels from A to B and back along route 3 such that the outward signal travels backwards and the return signal travels forwards in time; the plane of transmission would now be tilted upwards to intercept C at Z; the Newtonian plane of simultaneity for the reflection event P is now $t = -1$; the events X, L, and P lie on this plane. In this case and in case 2, the angle of the tilt for the plane of transmission is clearly arbitrary, given the fact that we may select any finite speed for the transmission signal.[15]

Assuming that we can indeed construct such planes of transmission consistently, it is clear that the speed and time direction of signals will vary continuously with spatial direction; for example, signals coming directly out of the page will have an infinite speed, since tilting the plane down or up does not affect the slope of the plane at right angles to A so that along this line the transmission plane coincides with the plane of simultaneity. Which direction is chosen for the instantaneous one-way speed is not an absolute characteristic of the spacetime, since it would clearly be a conventional matter which spatial direction is chosen as a reference for any tilts applied.

Therefore, even if light or some other signal appears to travel instantaneously on such round-trips, we need not select t = 0 as *the* plane of simultaneity. All points on the world line of a distant observer (i.e. on the path traced out in spacetime by that observer) are 'topologically simultaneous' with any given local event. And so we may select *any* event on this world line as 'the metrically simultaneous' event. Hence, the degree of uncertainty about which event may be said to be metrically simultaneous in this case becomes *far* greater than that in the STR case above.[16] It is indeed a matter of fact that light travels around closed spatial paths at a finite speed. However, we may not jump to the conclusion that instantaneous signals *must* imply absolute simultaneity. If we accept the arguments for conventionality in STR, then there seems little reason to resist the demand for conventions even in 'Newtonian' contexts where some signals may make round-trips in an instant.

SLOW CLOCK TRANSPORT

Could we side-step the problem by finding some other standard, free from any conventional element, which would allow us to determine the one-way speed of light, at least in the context of STR? Instead of using standard signal synchrony to determine simultaneity between distant events, we might use the method of slow transport of clocks. This method is recommended by Ellis and Bowman and by many others as the most promising convention-free approach to the synchronisation of distant events.[17] Although slow clock transport is set firmly within the context of STR, the general idea involved may be found in a renowned seventeenth-century attempt to measure the one-way speed of light. In the 1670s, the Danish astronomer Ole Rømer argued that the speed of light is finite and approximately 210,000 km/sec. Most scientists in the seventeenth century were reluctant to disagree with Descartes's verdict, given in 1637, that 'light can extend its rays instantaneously from the Sun to us'.[18] However, Rømer realised that the periods between the eclipses of Jupiter's moons ought to exhibit the same kind of regularity as other astronomical events of this kind; but this is not the case. He believed that the irregularities could be explained by the fact that our observations of the eclipses take place from different viewpoints: sometimes the Earth is relatively close to Jupiter and sometimes the Sun lies between the two planets. Hence, light from the eclipses needs to travel across the diameter of the Earth's orbit when the Earth is at its most distant. But when the Earth is at its closest to Jupiter the distance travelled by the light is clearly so much shorter. If the speed of light were infinitely fast, then the irregularities would remain a puzzle. But Rømer's calculations indicated that, in order to compensate for the observed irregularities, light should take some 22 minutes to travel across the diameter of the Earth's orbit.[19] Rømer then used the available estimate of the diameter to estimate the speed of light.

Rømer used clocks on Earth as the standard for his measurements. But these clocks are not stationary with respect to Jupiter and its moons. As we saw in Chapter 2, according to STR, the time registered by a clock as it moves between two given points in a given frame depends upon its velocity relative to the frame and on the path taken. Therefore we cannot simply assume that the journey the clocks make from one side of the Earth's orbit to the other has no effect whatsoever on the measurements of the times between eclipses. These measurements, of course, provide the data for the calculation of the speed of light. If the clocks' readings are affected by the journey around the Sun, we will need to make a suitable correction to compensate for these effects. The figure calculated for the speed of light will then be dependent upon the assumptions supporting this correction. But are these assumptions free from convention? May we use clocks which are transported from one location to another as non-conventional, objective standards for our measurements of the speed of light? Ellis and Bowman believe that we may use a moving clock to provide an objective figure for the speed of light so long as the clock is transported, step by step through tiny, equal intervals, at an extremely low speed relative to the frame in which the beginning and end of its journey are situated, i.e. by using the slow clock method. Although two initially synchronised clocks moving from A to B via different paths will in general not be synchronised at B, they will be synchronised if the relative velocity and therefore the speed of transportation is infinitesimally small. Ellis and Bowman argue that slow clock transport is possible; they remind us that the Lorentz transformations (LT) allow us to calculate what happens when clocks are moved from place to place. We do not actually need to slow moving clocks down to a stop; we merely need to be aware that the mathematical characteristics of the LT allow us to infer the possibility of synchronisation on the basis of the behaviour of actual clocks as we move them at slower and slower speeds. We may then correct any readings made by a real moving clock on this basis. However, others, following Grunbaum, argue that our use of any such correction depends upon a conventional decision to adopt a metric for STR spacetime which allows us to specify the timelike interval between events and to give sense to the idea of 'equal intervals'; they then argue that nothing compels us to choose this *particular* metric except our conventions.[20] Hence, they say, the calculation for the one-way speed of light still depends upon our conventional choice, rather than upon some objective feature of the world.

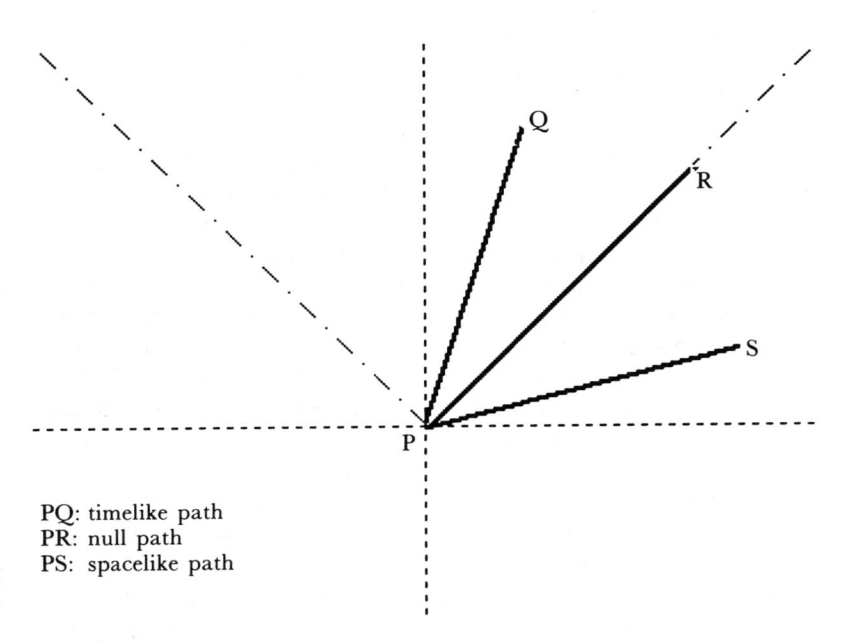

PQ: timelike path
PR: null path
PS: spacelike path

Figure 13 Pathways in spacetime

TIME, SPACE AND PHILOSOPHY

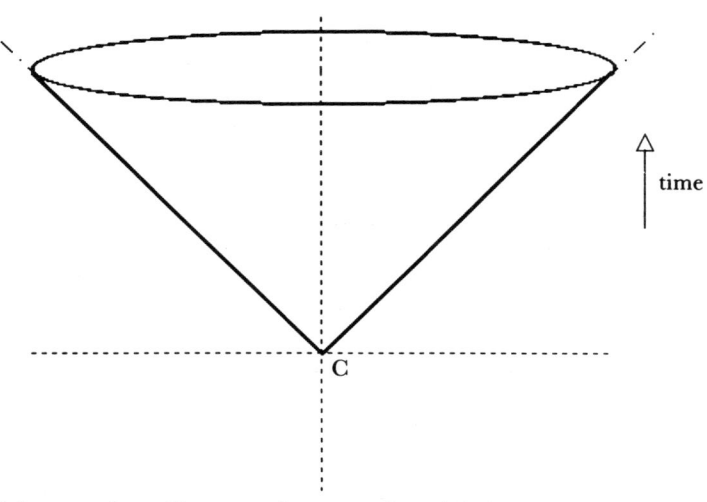

The light cone above illustrates the range of possible influence of the event C within two spatial dimensions (left/right; in/out).

Figure 14 Light cone

SPACELIKE TRAVEL: A TALE OF TWO TACHYONS

In the spacetimes of the STR and The General Theory of Relativity (GTR), we may connect any two points *geometrically* by timelike, null, or spacelike paths; see Figure 13 (p. 59). When a timelike path connects two points, a signal travelling at less than the speed of light may pass continuously from one point to the other. When a null path connects two points, only a signal travelling at the speed of light may pass between them. The notion of causality in relativity is tied to the idea of light being the fastest possible causal signal between two points. When an event C is said to be a cause of another event E, we usually say that these two events may be connected by non-spacelike curves, i.e. by either null or timelike paths, so that a signal travelling at or below the speed of light may travel between C and E. So the 'light cone' of an event C may be thought of as the cone of possible causal influence – C may be regarded as a physical cause of any event within or on the light cone; see Figure 14 above. However, spacelike paths are not confined to the light cone, so they are not regarded as possible *physical* pathways for causal signals: if nothing can travel faster than light, then nothing can follow a spacelike path. Hence, the idea of a spacelike path or curve may be regarded as merely an artefact which allows us to explore the general geometrical properties of spacetime. This is a standard view of relativity, which excludes the possibility of particles moving along spacelike curves.[21]

What happens when we relax the stipulation that no physical signal or particle may travel faster than light? This does not 'break the laws of relativity': the laws of STR do not legislate against particles travelling faster than light. The key idea behind STR is the *invariance* rather than the limiting character of the speed of light. The standard view of relativity incorporates the idea of the speed of light as maximal as an additional 'physically plausible' hypothesis. As a number of relativists have shown, we can construct frameworks consistent with STR in which particles may travel at 'super-luminal' velocities.[22] But, if we give such particles a place within the context of STR, we must face the rather unsettling consequences. Particles travelling faster than light are typically called 'tachyons'.[23] And tachyons, some writers tell us, may be used to transmit messages backwards in time! The tachyon may travel, we are often told, outside the light cone of the event which is regarded as its source. Hence, tachyons may follow spacelike paths.

Tachyons need not challenge the postulate of the invariance of the velocity of light. For, although the idea that light is the fastest signal is intimately connected with relativity, the metrical geometry of spacetimes in STR and GTR depends upon the distinct and fundamental notion of the invariance of the velocity of light. Hence, the framework in which we shall explore the behaviour of tachyons is constructed within the context of the postulate: light has the same invariant velocity in all inertial reference frames. This postulate allows us to define planes of simultaneity in spacetime. And this will in turn enable us to note any causal anomalies arising from the introduction of tachyons into the spacetime.

Although we shall see that there may be grounds for scepticism about the alleged causal anomalies associated with tachyons, the behaviour of tachyons does raise an interesting general point concerning attempts to redescribe backwards causation in terms of forwards causation. Many stories are told about tachyons, but most share the same general characteristics as revealed in the account given by R.C. Tolman and developed by David Bohm and others.[24] Imagine two instruments which are capable of emitting and absorbing tachyons. One instrument is close at hand in a laboratory L; the other is in a spaceship M moving away from us at a steady speed. We must note that, although we can represent this situation on a single spacetime diagram, we cannot draw a single plane of simultaneity common to both L and M. And, given the constraints of such diagrams, the world line of any object emitted at a steady speed from a source must depend upon the velocity of that source. The plane of simultaneity for a given observer (e.g. a scientist in the laboratory) represents the set of world lines of those signals moving infinitely fast away from that observer. Hence, distant events simultaneous with events in the laboratory L lie on a plane S_1 orthogonal to the laboratory's world line; but the plane of simultaneity S_2 of the spaceship M is tilted as shown in Figure 15 (p. 63). The faster a signal moves from a given observer, the 'closer' the world line of that signal will be to the observer's plane of simultaneity.

We now imagine an extremely fast tachyon moving away from L and towards M, represented by the world line P_1 (close to S_1). Notice that S_2 meets the world line of L at a time earlier than the emission of this tachyon; this is simply a result of the fact that two observers in relative motion will in general disagree about which events are simultaneous. The tachyon is eventually absorbed by the instrument on board M. This instrument is programmed to emit a tachyon shortly after absorbing a tachyon. So a further tachyon is then emitted towards L, at the same tremendous velocity and therefore close to its plane of simultaneity S_2, along the world line P_2. This second tachyon travels forwards in time with respect to the spaceship. But, given the tilt to the spaceship's plane of simultaneity, the tachyon meets the world line of L at a time earlier than the initial emission from L! Hence, from the laboratory point of view, the second tachyon seems to be travelling backwards in time, given the 'facts' that:

1 a tachyon follows the path P_1 leaving L at time t^*;
2 a tachyon follows the path P_2, meeting L at time t, such that t is earlier than t^*;
3 the tachyon following P_1 is absorbed by the instrument on the spaceship S; and the tachyon following P_2 is then emitted by this instrument.

If this story is coherent, then tachyons seem to allow the possibility of sending messages backwards in time. Of course, since people are

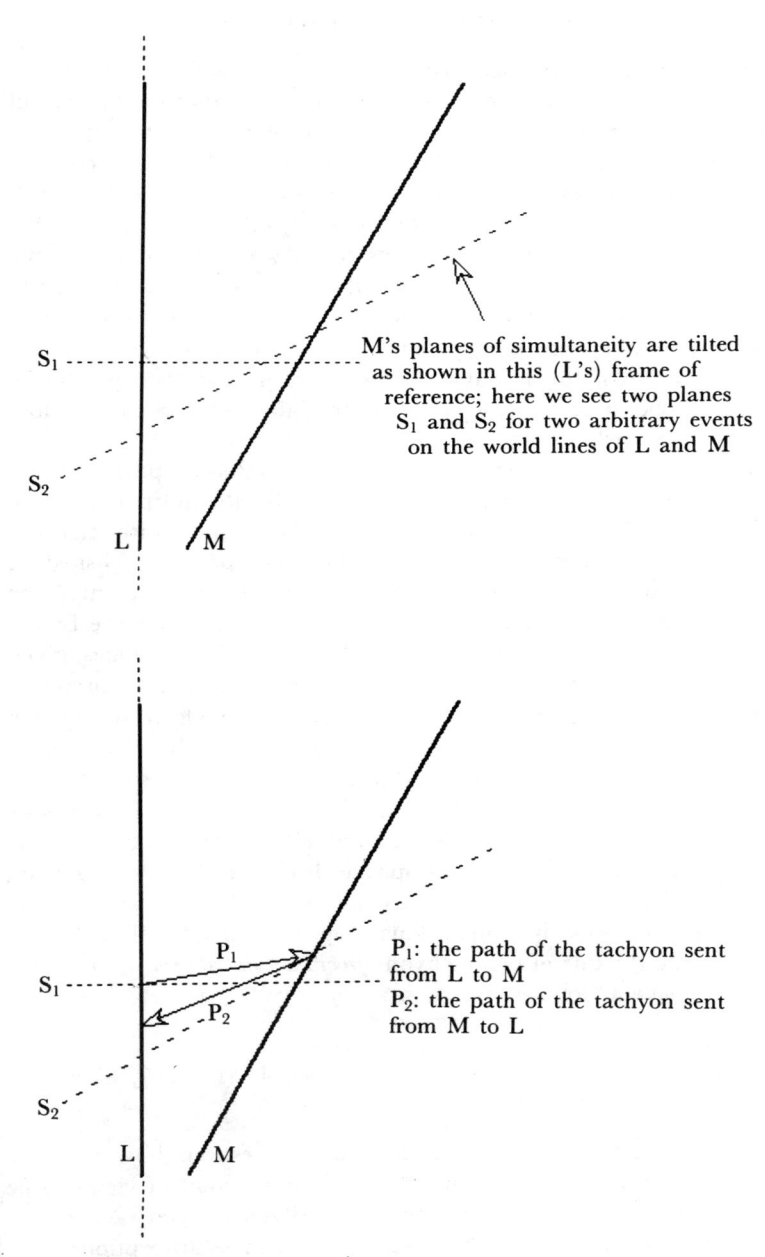

M's planes of simultaneity are tilted as shown in this (L's) frame of reference; here we see two planes S_1 and S_2 for two arbitrary events on the world lines of L and M

P_1: the path of the tachyon sent from L to M
P_2: the path of the tachyon sent from M to L

Figure 15 Tachyons: travel backwards in time?

clearly not made up of tachyons, there is no question of any human travelling backwards in this way. But there is still a serious causal anomaly here: the instrument in the laboratory seems to receive, at a given time, a signal which may be traced 'back' to an event at a later time in the laboratory. It seems that in such a case 'backwards causation' is possible: that is, an effect may in some circumstances happen before its cause. If we ask what causes the second tachyon to reach the laboratory, then the answer seems to be its emission from the spaceship; and this event only takes place because of the absorption of the first tachyon, which itself is a consequence of the initial emission from the laboratory. Hence, from the point of view of anyone in the laboratory, the effect does seem to precede the cause.

Particle physicists frequently refer to pairs of particles and antiparticles. The electron is paired with its antiparticle, the positron; similarly, the tachyon may be paired with the antitachyon. Richard Feynman claimed, in a paper published in 1949, that a positron travelling forwards in time may be reinterpreted as an electron moving backwards in time; see Figure 16 (p. 65).[25] This interpretation has a number of advantages, given that it allows physicists to consider two distant but correlated events as the effects of a single particle, an electron, moving forwards and backwards in time. We shall look at Feynman's idea in greater detail in Chapter 8. But, at this stage, we may note the general point arising from his analysis: a particle travelling in one direction in time may be *treated* as (if not identified with) its antiparticle travelling in the opposite direction. Hence, a tachyon moving backwards in time may be treated as an antitachyon moving forwards in time. This approach might allow us to redescribe the tale of two tachyons *from the point of view of someone in the laboratory* as follows:

1 a tachyon follows the path P_1 leaving L at time t^*;
2 an *antitachyon* follows the path P_2 *leaving* L at time t, where t is earlier than t^*.

This redescription raises an immediate question: how may we account for what seems (from the laboratory point of view) to be the near simultaneous absorption of antitachyon and tachyon by the instrument on board the spaceship? For these absorptions seem to demand a correlation between the emissions from the laboratory which may be hard to explain in any straightforward way. Is the

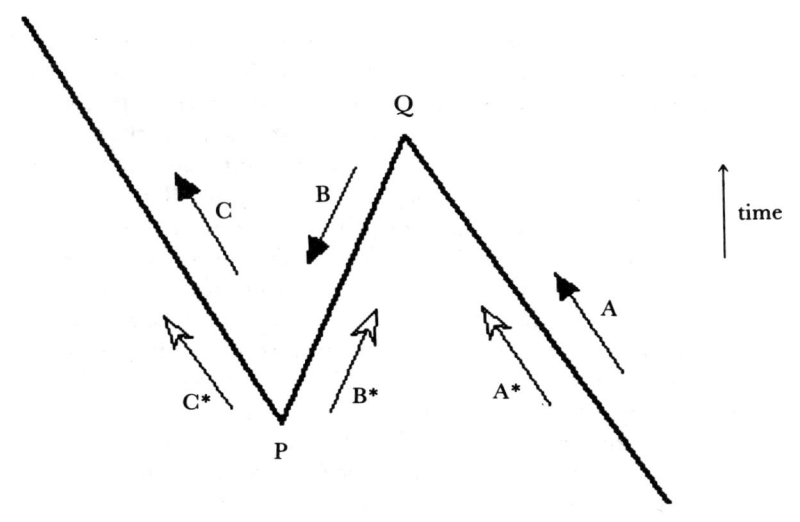

Feynman argues that we can tell stories about particles in different ways: consider the above situation: A* is the path of an electron; B* is the path of a positron: and C* is the path of another electron. P is the point where the electron following path C* and the positron following B* are created; and Q is the point where the positron and the electron following A* meet and annihilate each other. This story requires the ideas of particle creation and annihilation, and modern particle physics does indeed give substance to these ideas. But we can retell the story in terms of a single particle, an electron, which follows the route A → B → C. Of course we now need to provide some explanation of why the particle switches direction in time at P and Q. But we no longer need to lean on the ideas of creation and annihilation. Each story has its complexities and simplicities. The advantage of the second story is that we may treat positrons as electrons (in reverse time direction admittedly) – but we only need one kind of particle. In general, we might treat all antiparticles as particles travelling backwards in time.

Figure 16 Feynman diagram

laboratory instrument 'forced' to send out the tachyon having already emitted the antitachyon? If this is the case, why should it be so? These questions raise a general difficulty for many redescriptions of backwards travel and causation: do we not commit ourselves to unexplained correlations between events? This problem needs further exploration, and in Chapter 8 we shall examine the issues involved.

There are, however, strong grounds for suspicion about the

coherence of the tachyon story. To be consistent we must account for the fact that the instrument on board S is programmed to emit following an absorption. But this instrument does not seem to absorb anything at all; rather, as Recami shows, from the point of view of an observer on board S, it seems to emit two antitachyons! In a thorough analysis of the problem of tachyons, Recami argues that, for the instrument to behave as planned, the particle absorbed must leave the laboratory before the emitted particle returns. He goes on to suggest that even the idea of absorption and emission may be frame-dependent, so that what counts as an emission to one observer may be an absorption from a different point of view. The details of this analysis are complex, but Recami draws the moral that confusion will always arise when we 'mix together the descriptions of one phenomenon yielded by different observers' – and that the events on board S must not be described as 'fact' from the laboratory point of view.[26]

JUST THE TWO OF US: ACROSS THE UNIVERSE?

Robert Elliot suggests an ingenious way in which people and not just tachyons might travel 'faster than light'.[27] With judicious planning and by making use of advanced technology, he says that anyone may travel faster than light, i.e. follow a spacelike path. The (as yet unavailable) technological requirements are: the ability to manufacture a 'total' biochemical blueprint of the physiological and psychological composition of a given individual; and the ability to reconstruct an individual from the details of such a blueprint even after the transmission of these details to distant star systems. Elliot's plan for faster-than-light travel may be retold as follows:

1 On the Earth, a total blueprint is made of Edith – and this is stored safely.
2 Edith gives instructions that she should be embodied on Earth from the details of the blueprint after a precisely given time – say exactly ten years.
3 A copy of the blueprint is then sent by 'conventional' means to a planet in a distant star system at the speed of light – a journey which takes four years.
4 At the moment of transmission from the Earth, Edith is destroyed; but, when the blueprint reaches the planet, Edith is

reconstructed from the blueprint details.
5 After spending six years seeing the local sights, Edith is destroyed.
6 At the moment that Edith is destroyed on the distant planet, Edith is reconstructed from the blueprint details left on Earth.
7 The 'journey' from the planet to the Earth takes place instantaneously: at one moment, Edith is alive and well on the distant planet, and in an instant, she is relocated to Earth; so Edith travels faster than light!

Even if we accept that causal signals may travel no faster than light, Elliot assures us that anyone who carries out Edith's plan does indeed make an instantaneous change of spatial position such that only a spacelike curve may connect the points of departure and arrival. Of course we must accept two basic presumptions: first, it must be possible in principle to make such a blueprint from which an individual could be genuinely reconstructed; and, secondly, we must agree that a plane of simultaneity may be defined such that events on the distant planet and on the Earth may be said to be simultaneous. Let us grant both of these. However, the question of whether or not Edith does travel faster than light seems to turn on the claim that the Edith who left the star is 'psychologically continuous and cohesive' with the Edith reconstructed on the Earth.[28] If we agree that the first conventional 'journey', from Earth to planet, does not disrupt Edith's psychological and physiological continuity and cohesiveness, then Elliot says that there is little reason for us to object in the case of the second more problematic 'journey'. Let us also accept (if only cautiously) the assertion that personal identity may be preserved in cases like the first journey.[29] We might justify this decision by agreeing to Leibniz's principle of identity of indiscernibles: if two objects are identical in all respects, then they are one and the same object. Hence, if the Edith who appears on Earth is identical in all respects with the Edith who disappears from the distant planet, then we must be talking about just one Edith. So is Edith's 'return' to the Earth really a case of travel faster than light?

When Edith spends six years in the neighbourhood of the distant star, we would expect her to undergo many physiological and psychological changes during this time. She might fall in love or develop a fear of spiders; she might lose a tooth or acquire an ulcer. Consequently, the physiological and psychological composition of

Edith after six years is likely to differ in many respects compared with the composition of Edith as recorded on the blueprint on Earth. Even if we retell the story so that Edith spends just fractions of a second in her embodiment, changes will occur in Edith's composition – neurons will fire and cells will die: if she spends any time at all embodied on the planet then this Edith will not be identical with the Edith embodied on Earth. But this means that we lose the psycho-physical continuity essential to Elliot's claim that in such circumstances Edith would have travelled faster than light.

The two Ediths would only be genuinely continuous if no time at all elapses, from her point of view, between her embodiment and destruction on the distant planet. So genuine continuity could only be guaranteed if Edith is not embodied at all on the planet. But this means that there is no question of anyone travelling from the planet back to Earth at all! No Edith is available to make any sort of journey. And, if she were available, then the possibility of changes in the composition of this Edith disrupts the essential continuity needed for the identification of her with the Edith reconstructed from the original blueprint on Earth. So, even if we make some rather courageous assumptions about the nature of personal identity, we still fail to provide a convincing case for spacelike travel. So, if she really wants to go places, Edith ought to stick to more conventional means of travel.

4

A CONVENTIONAL WORLD?

INTRODUCTION

The shortest distance between two points is a straight line. Two parallel lines will tend to stay parallel. The internal angles of a triangle add up to 180 degrees. Such statements belong to the apparently secure domain of high school mathematics. They have as their basis the axioms of Euclidean geometry. The more useful, but complex theorems of Euclidean geometry are typically deductions from the basic definitions, axioms, and postulates of the system.[1] Why do we accept these basic ideas: because they are self-evidently true, because they are inherently rational, or because they are intuitively correct? All these reasons have indeed been advanced. But, instead of committing ourselves so whole-heartedly to the building-blocks of Euclidean geometry as somehow necessarily true, we might just accept them because they provide us with an internally consistent system which seems to underwrite so much that we say about the physical world. Then, if we are faced with a choice between geometries, we can make decisions without feeling the need to justify our choice from first principles.

The choice used to be simple: Euclidean geometry or nothing. And why should anyone question these axioms when they seemed to be so in tune with the evidence of our eyes? To doubt Euclid was almost as great a heresy as atheism. However, mathematicians like to play innovative games and they sometimes have scant respect for even the most sacred of cows. And along with the spirit of free enquiry in the nineteenth century came mathematicians bold enough to challenge Euclid and talented enough to construct alternative geometrical systems, in which the three angles of triangles might not add up to 180 degrees and in which lines parallel at one location might indeed meet at another. Some regarded the new geometries with suspicion. Henri Poincaré suggested in *Science and Hypothesis* (1902) that 'Euclidean geometry ... has nothing to fear from fresh experiments' – he believed that Euclidean and non-Euclidean geometries are distinct, incompatible theories which may be equally supported by empirical evidence of observation and experiment.[2] With a few 'trifling' if complicating assumptions, we could retain the Euclidean world-view and still be consistent with the empirical evidence. We might then resist the adoption of any non-Euclidean theory – despite any advantages such a new outlook might seem to offer. Hence, the decision to adopt either geometrical perspective is essentially conventional; see the previous chapter for some introductory remarks about conventionalism (pp. 46–9).

Euclid was destined to fight one last battle some 2200 years after his death. With the development of the General Theory of Relativity (GTR) in 1915–16 came the realisation that non-Euclidean geometries might be more than mathematical fantasies. These new geometries might be far more appropriate for our physical descriptions of gravitational and other fields than the Euclidean view. The non-Euclidean approach of GTR did not convince everyone that the Euclidean perspective was altogether redundant. Hans Reichenbach argued that the empirical evidence in favour of GTR need not force us to accept a non-Euclidean view. But, unlike Poincaré, he did not believe that two views which are equally supported by the evidence are genuinely distinct.[3] Reichenbach says that the non-Euclidean view is *empirically equivalent* to a Euclidean perspective, once we make additional assumptions about the physical world.[4] Of course, we might nevertheless have good (non-empirical) reasons for deciding to adopt a non-Euclidean view. But if we move away from a Euclidean view, the range of non-Euclidean geometries on offer might seem to be bewildering: spacetimes with metric fields, with scalar fields, and with vector, tensor, or non-dynamical fields. Indeed, some theories no longer restrict themselves to the four familiar dimensions of space and time, for example: the 'string' theories proposed by Green and Schwartz which employ multiple (and not all strictly spatio-temporal) dimensions; or the 'chaotic inflation' cosmological theory of Linde which allows for a compartmentalized universe – broken up into many distinct and isolated regions, some of which may have, say, just two spatial dimensions.[5]

A CONVENTIONAL WORLD?

In this chapter, we shall examine a variety of situations which might be said to involve either empirical equivalence or conventions. Our discussion of Euclidean versus non-Euclidean geometry will begin by focusing on the metric. We shall then investigate the problems raised by the topological structure of space and time. We shall then try to get a firmer grip on the general philosophical problems raised by Reichenbach and Poincaré; but sadly this is as far as we can take the discussion of conventionalism in this book. The issues are complex, and the aim here is to clarify the central problems as much as possible. But simply trying to reach this goal will bring many rewards: for we shall also obtain a more comprehensive picture of both Euclidean and non-Euclidean geometries and also of relativity theory.[6]

WHEN PARALLEL LINES MEET

The first 'convention' which we shall examine is the choice between Euclidean and non-Euclidean *metrical* geometry. The geometrical postulate which caused all the trouble for the Euclidean picture is the fifth, the postulate of the parallels, which states that:

> through any given point there is one and only one parallel to a given straight line (which does not go through the given point), i.e. one straight line which lies in the same plane with the first and does not intersect it.
> (Reichenbach 1957: 2)

Figure 17 (p. 72) illustrates the situation envisaged by the postulate. Only one line through P is possible if that line is not to intersect a straight line AB whilst remaining in the same plane as AB. We may sympathise with those who felt that this postulate is self-evidently true; but, as Reichenbach remarks, perhaps we should be suspicious about the status of the postulate since the prohibition against intersection seems to continue beyond the confines of our finite experience right on to infinity. Carl Friedrich Gauss, Janos Bolyai, and Nikolai Ivanovich Lobatschevskii are believed to be jointly responsible for the challenge to this axiom, thus producing the first non-Euclidean geometry; only the last two mathematicians published their results – in the late 1820s and early 1830s – and the new geometry was shown to be internally consistent some forty years later by Eugenio Beltrami.[7] The end

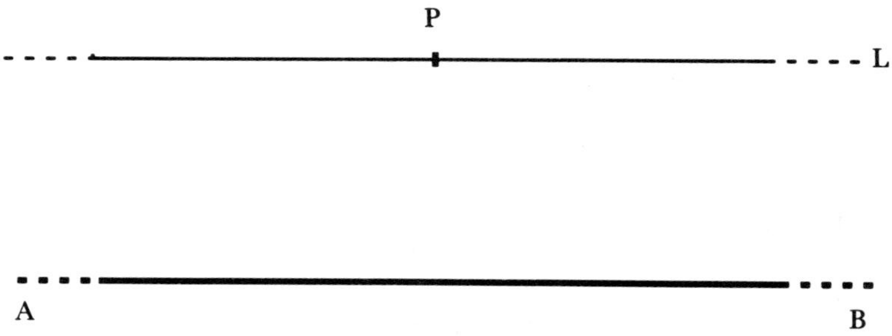

```
                              P
- - - - ●────────────────────●───────────────────── - - - - L



  - - - ●───────────────────────────────────────● - - -
  A                                              B
```

Any line other than L will intersect the continuation of line AB according to Euclid's parallel postulate. Line L is said to be parallel to AB, and it is the only parallel line which may be drawn through P in the same plane as AB.

Figure 17 Parallel postulate of Euclidean geometry

result of their combined efforts is a geometrical model which is consistent with all of Euclid's axioms except the fifth. In this 'non-Euclidean' model, there may be more than one parallel line through P. The three angles of a triangle will add up to *less* than 180 degrees. Although the parallel lines and the sides of the triangle are not 'straight' in a Euclidean sense, they are nevertheless as straight as possible within this non-Euclidean model: such lines are called 'geodesics'. Further work by Georg Riemann led the mathematical community towards the idea of generally curved spaces in which curvature might vary arbitrarily from point to point so that space might be 'flat' or Euclidean at one point and 'curved' at another.[8] Riemann's work also suggested the idea of a positively curved 'spherical' space, which is perhaps the most easily imagined non-Euclidean space: see Figures 18 and 19 (pp. 73, 74); and his mathematical genius contributed towards the development of the tensor calculus which provided the mathematical tools needed by Einstein and his co-workers in their construction of GTR.

Some mathematicians, including Riemann, did ask which kind of geometry should be employed to describe the observed physical world. However, without firm empirical evidence, there was very little chance of non-Euclidean geometries being regarded as anything more than mathematical curiosities, tributes to the tremendous ingenuity of their discoverers. The development of the

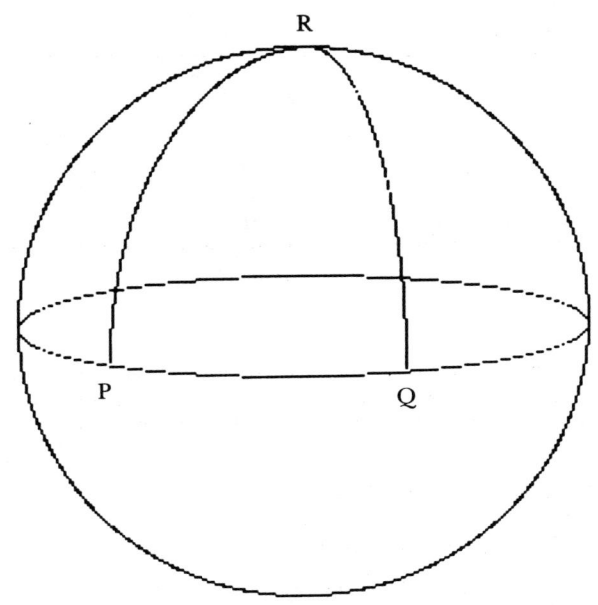

The three internal angles of the triangle PQR on the surface of the sphere above add up to more than 180 degrees. Two parallel lines through P and Q heading directly 'north' will meet at the 'pole' of the sphere.

Figure 18 Idea of curvature on surface of a sphere

Special Theory of Relativity (STR) and then GTR challenged this situation. STR involves a four-dimensional view of the physical world. Minkowski shows that the most natural way to describe physical reality is in terms of events in a four-dimensional framework: spacetime.[9] And GTR links the distribution of matter and energy throughout the physical universe to the metrical geometry of spacetime. This link highlights the relationship between motion and curvature; in GTR, as the eminent theoretical physicist John Archibald Wheeler so elegantly says: matter tells spacetime how to curve, and spacetime tells matter how to move.[10] Indeed, even freely moving massless particles, such as light, follow the geodesics of spacetime in GTR. And these geodesics are determined in part by the distribution of matter. A massive object, such as the Sun, causes marked curvature in its immediate neighbourhood; see Figure 20 (p. 75). Hence GTR predicts that

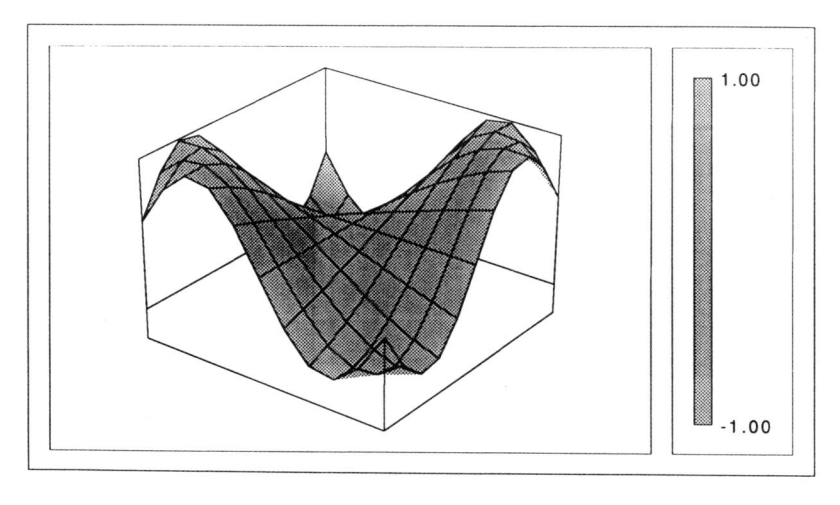

This illustration presents the essential idea of a generally curved spacetime. Here we see a two-dimensional surface curved. The grid on the surface represents the paths which would be taken by free particles or rays of light. But the analogy between the curvature of spacetime and that of this surface is limited. The two-dimensional surface sits in a three-dimensional box. However, there is no higher spatial dimension in which three-dimensional space sits. Spacetime is all we have, and it is a mistake to think in terms of spacetime as a three-dimensional spatial structure sitting in some background container.

Figure 19 Idea of a generally curved spacetime

light from any distant star follows a curved path as the light passes close by the Sun. Physicists are normally reluctant to embrace any theory without some hard empirical evidence. The confirmation of this prediction, by the astronomer Sir Arthur Eddington in 1919, helped to convince most of them that we inhabit a non-Euclidean world.[11]

WILL THE REAL GEOMETRY PLEASE STAND UP?

The empirical evidence points towards a non-Euclidean world. However, if the message of conventionalism is correct, then we ought to be able to construct a theory which is the empirical

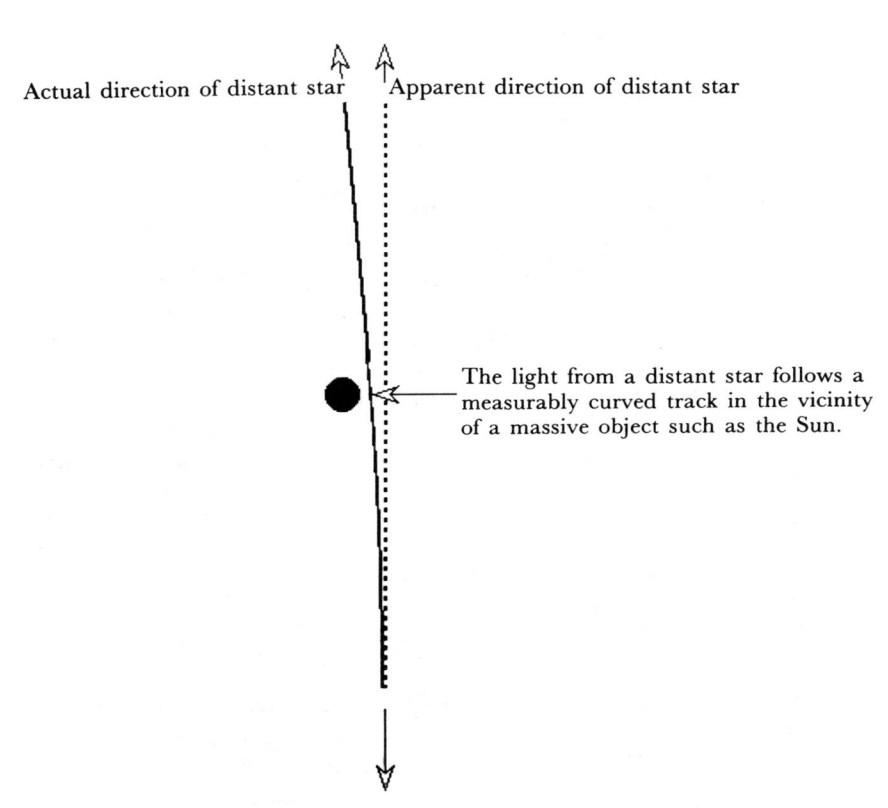

Actual direction of distant star | Apparent direction of distant star

The light from a distant star follows a measurably curved track in the vicinity of a massive object such as the Sun.

Observers see light coming 'straight' at them from the star. Comparison of the position of the star before and when the Sun moves across the line of sight demonstrates that the light does not follow the same path when a massive object approaches that path. In certain circumstances, we can actually see behind the Sun!

Figure 20 Bending of light in gravitational field

equivalent of GTR, but which implies that the metrical geometry of the world is Euclidean despite the evidence to the contrary and which we can adopt without absurdity. Poincaré and Reichenbach suggest that such worlds are indeed possible and the tale which follows is based on their suggestions.[12]

Imagine that the inhabitants of one of the outer planets, Neptune, have always used Euclidean geometry and they are convinced that the universe is Euclidean: every measurement they have made seems to be in line with this geometry. They know of no other possible geometry and they have no reason to doubt the status of the Euclidean view. But they are curious about the inner planets, and a party from Neptune sets off in a spaceship to explore the solar system closer to the Sun. On their journey, in order to keep the spaceship on course, they make a series of measurements with a variety of measuring instruments; some of these are basic instruments such as metre rules and clocks, and others are more sophisticated such as telescopes. These instruments have common standards of spatial distance and temporal duration. All goes well until the craft passes inside the orbit of Earth. They then notice that every measurement they make no longer conforms to the Euclidean view. For example, the light from distant stars now seems to be following curved rather than straight paths; and the measurements made on triangles within the spaceship show small but significant deviations from expected values. Wherever they explore within the sphere defined by the Earth's orbit, the story is the same. And the results they obtain are the same regardless of the kind of measuring instrument used. However, as soon as they move outside this orbit, their measurements conform to the Euclidean view once again. They enter and leave the sphere defined by the orbit several times, checking and rechecking this strange phenomenon. As good Euclideans, they are reluctant to doubt the applicability of their long-standing geometrical perspective. In fact, it never even occurs to them that the Euclidean view might be wrong or in need of some correction. Instead they decide that there must be some general and all-pervasive influence at work within the sphere defined by the Earth's orbit which distorts all their measuring instruments. They also conclude that this influence distorts everything, including the paths of light beams, which might be used to make any measurement of distance. Unlike other natural phenomena with which they are familiar, such as heat or magnetic forces, this strange influence affects all objects in exactly the same way. For example, copper and other metal triangles seem to be just as distorted as plastic or wood triangles.

They decide to fly back to Neptune to relate their startling discoveries. On their return, they find that equally startling advances in geometry have been made in their absence. A group of mathematicians have shown that it is possible to construct non-Euclidean geometries. A fierce debate then starts on Neptune. Some argue that they should retain the Euclidean view. They explain the curious measurements obtained by the explorers in terms of 'mysterious' universal forces which distort everything

A CONVENTIONAL WORLD?

within the Earth's orbit – after all, had not Neptunians always felt that Earth was an odd place?[13] Others claim that the discovery of non-Euclidean geometries solves the whole problem: outside the Earth's orbit, space is Euclidean; but inside it is non-Euclidean and deviations from the Euclidean view are to be expected.

So Neptunians must now make a choice between two alternative hypotheses about the space inside the Earth's orbit:

1 that it is Euclidean just like space elsewhere, but a universal force is in operation there which distorts everything including every kind of measuring instrument; and
2 that the solar system is Euclidean except for a central non-Euclidean 'irregularity'.

The second hypothesis forces the Neptunians to accept that they live in a universe which is, at least in part, non-Euclidean. So, if they adopt 2, Euclidean geometry alone is insufficient to describe their environment. However, the postulation of universal forces demanded by 1 allows them to retain a Euclidean account of the geometry of the universe. So is there any convincing reason why Neptunians should accept 2 and reject 1? Could we not say that the simpler view should be adopted? But the situation here is not so simple! For both hypotheses seem to require complicating rather than simplifying changes to the community's system of beliefs: 2 has its non-Euclidean geometrical complexities; and 1 involves an additional force with a local (if all-pervasive influence) in the central region of the solar system. And, because the universal force, if present, affects all objects in this region in the same way, we cannot *prove* its presence independently. So we cannot offer some independent empirical evidence in favour of 1. The evidence seems to support each hypothesis to the same degree.

It would also be dangerous to try to defend hypothesis 1 on the grounds that Neptunians (and humans for that matter) can only *visualise* the world in a Euclidean way. Clearly, there is no *logical* reason why they should be so restricted, especially given the fact that logically consistent non-Euclidean geometries have been constructed. Are we to say that such constructions do not involve 'visualisation' in some sense? As Reichenbach points out, they do not involve visualisation if we simply mean by that a peculiarly 'Euclidean visualisation'; he says: 'we cannot visualise non-Euclidean geometry by means of Euclidean elements of visualisation'.[14] But we can and do use Euclidean 'pictures' as analogies to help us understand non-Euclidean geometry. Despite the differences between the geometries, there are, of course, many similarities which provide the analogical base for us to do this – for example, the properties of 'straight lines' on the surface of a sphere discussed in Figure 18 (p. 73) provide a two-dimensional Euclidean analogue of a three-dimensional non-Euclidean configuration. However, a coherent visualisation of a non-Euclidean spacetime may only be achieved from a non-Euclidean perspective. Our difficulties in achieving this easily seem to arise from the fact that the Euclidean metrical perspective is deeply entrenched in our system of beliefs. It is hard to break free from such a habit. But the construction of non-Euclidean models demonstrates that it is indeed possible to break away.[15]

So Neptunians seem to be free to choose either 1 or 2. It might seem odd to choose a theoretical model which involves a strange and highly localised force. But why should this be any more peculiar than admitting that the geometry of the world has such an unusual anomaly? The choice seems to be between accepting either a physical or a geometrical mystery.

Of course, the tale about the Neptunians is fictional in more than one way. If a party of Neptunians were to investigate the geometry of the solar system, then they would not find any *sharp* transition between Euclidean space and apparently non-Euclidean space. Instead, if they were to approach the Sun, they would find that space becomes increasingly non-Euclidean – with very marked curvature close by the Sun. GTR gives a good explanation of this: the theory relates the non-Euclidean character of space (and time) to the distribution of mass and energy. A massive, dense body such as the Sun produces sufficient local curvature to be detectable quite readily with a range of measuring instruments. Nevertheless, the Neptunians could continue to hold their Euclidean beliefs even with this more accurate account of the kinds of measurements which would be made. They might still believe that some universal force influences all measurements. But GTR gives them a sound theoretical basis for a non-Euclidean view. The whole point about GTR is that it demystifies the presence of curvature: the mass (and energy) in a given location is related (by a set of equations) to the curvature in that region. Hence, the theoretical framework of GTR encourages us to accept the non-Euclidean hypothesis in our own universe. The choice is now between a mysterious force and a rather less mysterious if novel geometrical perspective. Reichenbach and Poincaré say that we *could* resist, complicate our account of physical forces, and maintain a Euclidean perspective. But the cost of this resistance is a theoretical concoction which seems clumsy and artificial rather than the more elegant approach of GTR.

CONVENTION AND TOPOLOGY

Before we specify the metrical geometry in a spacetime, there is no well-defined sense to the ideas of lengths, distances, or duration in that spacetime. But such a 'metrically amorphous' spacetime does have structure: we may specify its affine and conformal structure or its topological characteristics – its dimensions, connectedness, compactness, and orientability – independently of any metrical features; see Figure 11 (p. 54) in Chapter 3 for a general review of the different levels of structure in spacetime.[16]

An infinitely extended plane does not have the same topology as the surface of a cylinder. The plane is connected in a way that the cylinder is not: the fundamental topological difference between them lies in the fact that, on the plane, any closed path may be shrunk continuously down to a point and therefore the plane is 'simply connected'; but, on the surface of the cylinder, some closed paths may not be shrunk to a point in the same way – for example, those around the circumference.[17] The possibility of such topological differences suggests this question: are we always able to specify the topology of a spacetime on the basis of empirical evidence, or is there some element of conventional choice involved?[18] We have already encountered one topological dispute about the structure of spacetime, in our discussion of Zeno and the continuum.[19] The choice between continuous or merely dense orderings of the points in a spacetime seems to be essentially conventional. However, the idea of a merely dense three-dimensional space may not be theoretically coherent. So the more straightforward topological difference between the plane and the surface of a cylinder may help us to investigate the problem of conventionality within the context of topology with a little more clarity.

Many cosmologists believe that the approximate overall shape of the universe may be compared with an infinitely extended plane, so that any straight line starting at any given point will go on and on without end. In such a spacetime, a spacecraft moving away from

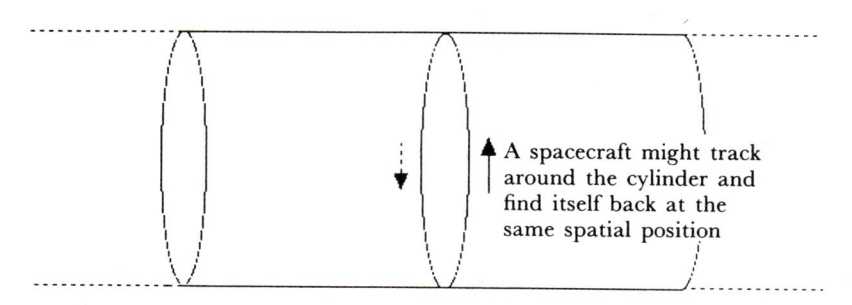

Figure 21(a) Topology of cylinder world

(Caption inside figure: A spacecraft might track around the cylinder and find itself back at the same spatial position)

the Earth would continue to do just that, with the distance between them growing ever-greater. Contrast this spacetime with a universe which possesses a cylindrical topology, in which straight lines may traverse the same ground by tracking around the circumference of the cylinder; see Figure 21(a) above. In this case, a spacecraft might travel away from the Earth moving around the cylinder and find its way back to the Earth without ever changing direction. Could we determine which kind of universe we inhabit? Clearly, the metrical structure of the two worlds might be different, since the cylindrical world might have a curved metric, whereas the plane is always metrically flat. But we might invoke the idea of universal forces to account for this metrical difference; or we might concern ourselves only with 'cylinders' in a topological sense only. But whatever we say, there is still an important topological difference between the two worlds: the plane is simply connected and the surface of the cylinder is not. And we might discover some empirical evidence associated with this difference. But now imagine an infinite plane which is divided into a series of strips, each of which is qualitatively the same; see Figure 21(b) (p. 81). What difference might we find between this segmented plane and a cylindrical world which shares the same qualitative features with any one strip? If we could journey around the cylinder, then we would see, over and over again, the same features and the same events. But, if we travel across the plane, each new strip encountered would present us with these same features and events. Despite the topological difference between the two worlds, it is hard to imagine what kind of empirical evidence might be unearthed to allow us to distinguish between these apparently identical domains.[20] Hence, it might seem that, once again, we may choose freely between two distinct worlds. However, as

Each successive strip has exactly the same features – instead of one Earth to which a spacecraft might return by tracking around the cylinder, there are now multiple Earths – one for each strip

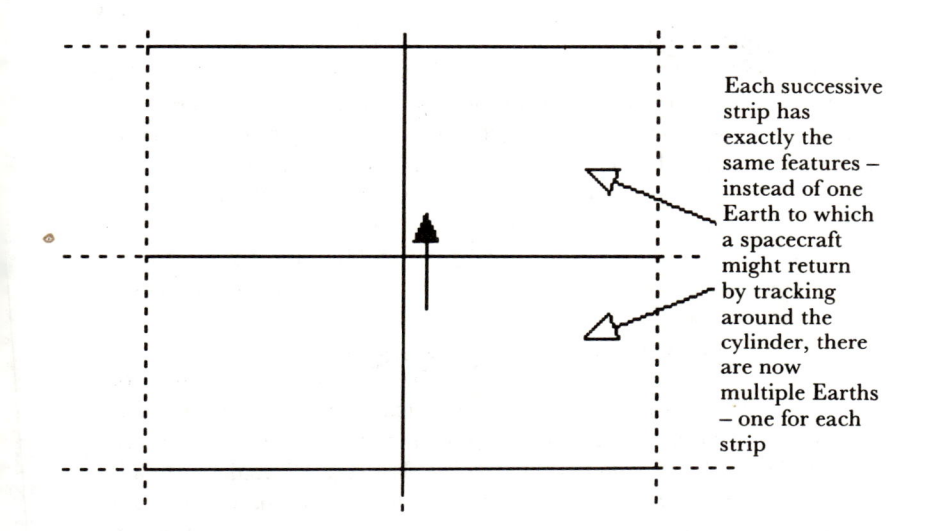

Figure 21(b) Topology of strip-world

Reichenbach advises, we must spell out all the consequences of competing views before making a decision in such cases.[21]

In the cylindrical world, every event is fixed by its unique position on the cylinder. However, in the strip-plane world, any single event has an infinite number of occurrences – one on each strip. The chain of events leading to any single event also repeats endlessly. So, if we resist the hypothesis that the world is cylindrical, we must say:

1 the world has the global topology of a plane;
2 universal forces are the cause of any observed deviations from the metric of an infinitely extended plane; and
3 an event which is causally fixed at one point must also be causally fixed in the same way on all other 'strips'.

Reichenbach says that beliefs such as the third lead to causal anomalies: 'The interdependence of all events at corresponding points cannot be interpreted as ordinary causality, because it does not require time for its transference and does not spread as a continuous effect that must pass consecutively through the intermediate points' (Reichenbach 1957: 65). If there is not some mysterious causal connection between each strip, then the only other explanation seems to be a cosmic coincidence or 'pre-established harmony' which underwrites this apparently miraculous repetition. Either way we would be choosing a strange world indeed if we decide to reject the more causally respectable cylindrical world. Yet even the cylindrical world has its causal problems – for its closed paths may be closed in time, leading to the controversial possibilities of time travel and of an event in my future influencing an event in my past. In Chapter 8 we shall take a closer look at such possibilities and the causal issues involved.

DIMENSIONS

One of the central features of Euclidean geometry set in a temporal framework is its 'three-plus-one'-dimensionality: forward-back; left-right; up-down; before-after. It seems difficult to imagine that we are wrong about the four dimensions of space and time. But are the worlds with different dimensionality *physically* possible? Riemann showed us that we may construct consistent multi-dimensional worlds: this formed the core of his work on non-Euclidean geometries. When we think of multi-dimensional worlds, we regard dimension, not as a 'physical property', but as a 'degree of freedom' or as a 'variable' needed to describe a topological manifold. A multi-dimensional manifold is the analogue of the two-dimensional surface of a sphere. We need two variables to pick out any point on the surface of a sphere: e.g. east-west and north-south references are all we need to pick out precise locations on the surface of the Earth. In a three-dimensional space, we need three variables. In an N-dimensional space, we need N variables.

After the development of GTR, several physicists investigated the relationship between dimensionality and the physical world. Ehrenfest argued that many physical stabilities are direct consequences of a universe with only three spatial dimensions; and Weyl showed that conformal properties would be invariant only in such a universe.[22] But the success of the geometricisation of gravity by Einstein led many to attempt the geometricisation of all material properties. Physics would stick with three spatial dimensions, but, in addition to a further dimension for time, there came a growing demand for extra dimensions to characterise the properties of matter. Superstring theory is the latest version of this programme. This theory, developed by John Schwarz and Michael Green, involves an additional six dimensions in a 'string' tightly wrapped up into a ball in a space only 10^{-35} metres across. These dimensions allow us to represent the various properties of matter in terms of the various modes which the string may be in: just as real strings may oscillate and vibrate, these superstrings have various modes of 'vibration' and each mode represents a particular property of matter.[23] Some physicists welcome this highly speculative theory, but others remain deeply sceptical. Richard Feynman's typically caustic response, made shortly before he died, reveals the antagonism of those who share his commitment to the need for a firm empirical basis before accepting a theory:

> I'm an old man now, and these are new ideas, and they look crazy to me, and they look like they are on the wrong track. Now I know other old men have been very foolish in saying things like this, and therefore, I would be very foolish to say this is nonsense. I am going to be very foolish, because I do feel strongly that this is nonsense! ... So perhaps I could entertain future historians by saying I think all this superstring stuff is crazy ... I don't like that for anything that disagrees with an experiment, they cook up an explanation ... it's a question of verifying your ideas against experiment.
>
> (Feynman 1988: 193–4)[24]

But, as the theory is developed and its theoretical power grows, the empirical basis may also begin to grow.

Although superstring theory is speculative, it resists the temptation to add or subtract spatial dimensions. Andrei Linde is less cautious. The 'chaotic inflation' cosmological theory of Linde allows for a compartmentalized universe – broken up into distinct and isolated regions. And some of these regions may have fewer than three spatial dimensions.[25] If we grant Linde's idea as a physical possibility, could the number of spatial dimensions be other than three in *our* region of the universe? John Barrow suggests that we are restricted to three dimensions by the stable atomic conditions required for life: 'The dimensionality of the universe is a reason for the existence of chemistry and therefore, most probably, for chemists also' (Barrow 1983: 339).[26]

Barrow's claim is underwritten, in part, by the Anthropic Principle – a principle which makes a decisive link between physics and biology. We shall explore this idea further in Chapter 9.

THE FUTURE OF THE UNIVERSE

Our own universe does not seem to have anything like a cylindrical structure – metrically or topologically. So it might be thought that such questions about the overall structure of the universe are somewhat contrived. But problems about global structure do face us. Is our universe closed or open? Or does the universe follow a middle way between these two possibilities – the so-called 'flat' universe? Or is the universe cyclical? What we decide has implications for the way we will see the future. For each of these four kinds of universe has a different future and, in addition, the cyclical universe involves a different past:

1 A 'closed' universe is usually characterised as one which expands from a singularity with extremely hot, dense initial conditions (the so-called 'big bang') and then collapses towards a final singularity.
2 An 'open' universe never stops expanding: the average overall density of matter in such a universe is not large enough to stop the material in the universe getting further and further apart. Gravitational attraction tends to slow the expansion down, but in an open universe the overall density of the material contents is insufficient to counteract the global expansion.
3 A 'flat' universe holds the balance between a closed and an open universe; there is a natural limit to the global expansion: the overall density is just sufficient to bring the expansion to a halt 'at infinity'.
4 A cyclical universe is one which follows a 'closed' pattern of expansion and collapse, but which 'bounces' out of the final singularity just before the moment of total collapse to nothingness to start the process all over again allowing the cycle to repeat endlessly – this phase of the universe is just one of an infinite number of cycles; see Figure 22 (p. 85).

As in the case of the cylindrical and plane worlds, the closed and cyclical universes have distinct metrics and topologies. Even though the 'local' metrical and topological features of a given cycle of expansion and collapse in the cyclical universe may be the same as those in the closed universe, there are obvious global differences between them: for example, the cyclical universe is infinite in a way that the closed universe, with a more clearly defined 'beginning' and 'end', clearly is not; but see the discussion of the idea of a

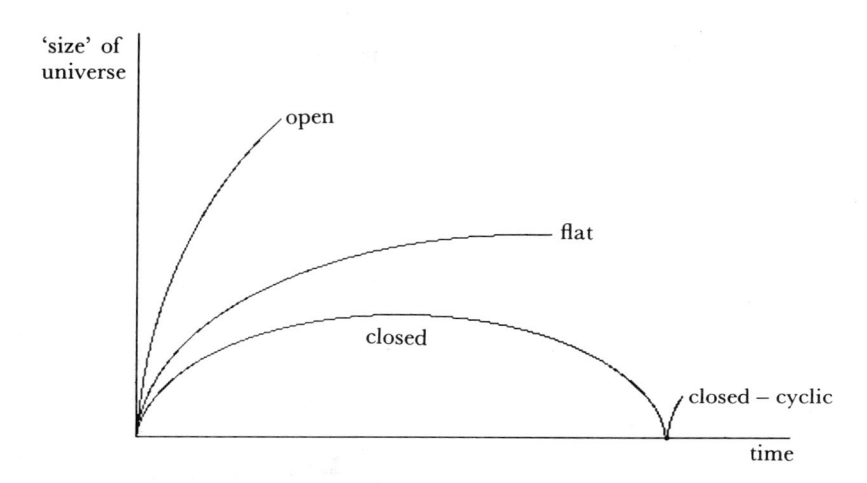

Figure 22 Graph showing open, closed, flat, and cyclic universes

beginning of the universe in Chapter 10.[27]

Whether we live in a closed universe or not is an 'open' question! The available evidence supports neither endless expansion nor some future recontraction – a 'flat' universe is the most likely alternative. But the important Friedmann cosmological models of the universe provide a cogent theoretical basis for each of the three main options: open, closed, or flat.[28] Additional evidence might indeed push us to adopt any one of these options: we might, for example, discover that the average density of the universe is more than sufficient to bring about global gravitational collapse. And even the most ardent conventionalist supporter of an open universe might begin to accept that the universe is closed when its material contents begin to collapse inwards. However, we might not be able to choose between a closed universe and a cyclical universe on *so* clear an empirical basis. Suppose that the evidence were to suggest that we live in a closed universe, how might we tell whether or not the expansion and collapse sequence happens only once? The conditions close to the initial and final singularities might obscure the existence of any bounce and therefore the possibility of another cycle, leaving us uncertain as to the eventual fate of the universe. So we might be tempted to allow the possibility of a cyclical universe – perhaps on conventional grounds. Of course, if we favour a cyclical view, we could not say whether each cycle would be identical in all respects to the observed expansion-recollapse pattern of the closed universe. Indeed, there is some evidence which suggests that each cycle would *have* to be different, for example, at lower and lower entropy levels.[29] However, it is hard to imagine what kind of physical principles might be involved in, and what sort of phenomena might be associated with, a 'bounce'.

Our knowledge of the future of the universe seems to be, at least at present, somewhat fragmentary when compared with our understanding of more local events and phenomena. So we are left in a (relative) state of ignorance with (hopefully informed) speculation as our guide. It seems clear that our inability to make a choice between futures seems to arise simply because of the *lack* of available evidence, rather than because decisive empirical evidence could not be found *in principle*.

THE COSMOLOGICAL PRINCIPLE: CONVENTION OR FACT?

Wherever we look, whatever direction our telescopes point, the universe seems to be much the same. Clusters of galaxies seem to be spread throughout the universe in a generally uniform way and there seem to be few large-scale anomalies. Hence, from our perspective, the universe appears on a *global* scale to be approximately:

1 the same at all points, i.e. homogeneous; and
2 the same in all directions, i.e. isotropic.

The central assumption in most accepted cosmological theories is that the observed large-scale distribution of matter and energy is so smooth that the universe is generally homogeneous and isotropic everywhere. Hence any observer anywhere would see essentially the same large-scale picture; see Figure 23 (p. 87). This assumption is called the Cosmological Principle and it lies at the heart of the standard Friedmann cosmological models of the universe. This 'Copernican' idea that neither the Earth nor its general environment has a privileged location in the cosmic 'scheme' is developed to deliver a truly egalitarian universe. Neither the solar system, nor our galaxy, nor any other galaxy, for that matter, is regarded as being 'central' or having any special status in the universe as a whole. The universe expands but need not have a 'centre' as such. Two-dimensional analogies may be

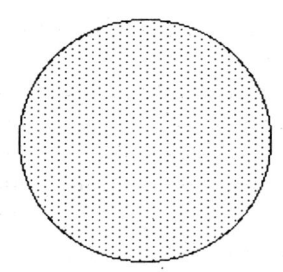

A repeating pattern on the surface of a sphere illustrates the idea of homogeneity – wherever we look we find essentially the same situation, the same local picture; and, regardless of the point we choose as a reference, essentially the same view presents itself in every direction, thus illustrating the idea of isotropy.

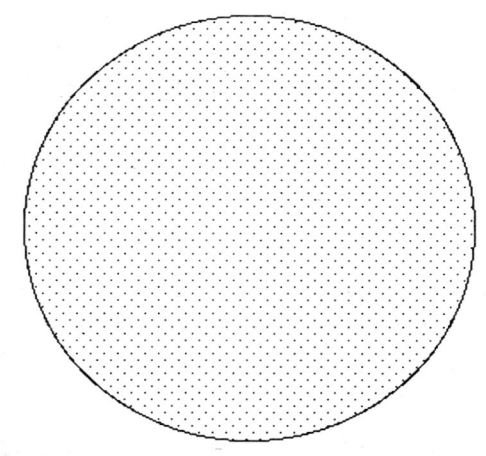

As the sphere increases in size, it remains homogeneous and isotropic; and exactly the same amount of material is present as in the earlier picture.

Figure 23 Idea of homogeneity: expanding sphere

used to illustrate this idea: the surface of an expanding sphere does not have a centre; nor does the surface of an expanding infinite plane; nor indeed does the spatially and temporally closed surface of a torus. All points and all directions on such surfaces are equivalent. And the most likely geometry for our spacetime seems to be the four-dimensional analogue of either a very gently curving sphere or an infinitely extended flat or nearly flat plane.[30]

It is rare to hear any serious and informed challenge to the empirical status of the Cosmological Principle. But such a challenge is made by the eminent physicist George Ellis in his prize-winning essay: 'Is the universe expanding?'[31] He argues that its place in cosmology is guaranteed more by philosophical commitment than by empirical evidence. For we may construct other universes which are consistent with the available evidence but which are nevertheless inhomogeneous and anisotropic. To prove his point, Ellis constructs a spacetime in which matter circulates between two 'centres' – it is pumped out from one and sucked in at the other; both of these 'centres' are singularities. The universe does not expand, but the impression of expansion is given by the continuous movement away from one singularity and towards the other. The conditions at the first singularity are identical with those at the initial singularity in standard cosmological models. Our own galaxy is in a preferred position in the sense that only in certain periods in the life of a galaxy do physical conditions permit the existence of people. As our galaxy moves towards the second singularity, conditions deteriorate and human life is no longer possible. But other galaxies are continuously moving towards the preferred position for life. So we can say, at least in principle, that human life might continue always in galaxies whilst they occupy this preferred position.

Ellis avoids making any over-enthusiastic claims for his model universe. He prefers to point to the model as a warning to those who are uncritical in their acceptance of the Cosmological Principle. Yes, this principle has an excellent empirical pedigree; but every observation in favour of the standard view also supports the Ellis model: hence the assertion that the universe is homogeneous and isotropic certainly seems to have a conventional character. Of course, we might say that the standard models are more respectable inasmuch as we have developed and tested them extensively. Ellis replies by asking why there is so much reluctance to step outside the standard view. If empirical evidence alone is to be the touchstone for acceptability, then there is no reason to prefer this view to that constructed by Ellis.

Ellis has also argued, with Ruth Williams, that homogeneity and isotropy may be explained more powerfully using a model sometimes referred to as a 'small' universe. At its simplest, such a universe has the topological spatially closed structure of a torus; see Figure 24 (p. 89).[32] An interesting feature of the small universe is that its spatial closure gives us access to *all* events in our past.

The torus has the topology of a ring: the topology of the small universe is that of the surface of such a ring.

Figure 24 Topology of torus

When we look into our past we see not an infinite world but a finite world with an infinite number of 'images'. Wherever we look, we see a huge collection of galaxies; but, as we look further and further away, we see multiple images of this same collection. Each image is the result of light travelling a finite number of times around the torus. If we look in a different direction, we see a different set of images, but these are still images of the same unique object – the finite collection of galaxies which constitutes the total material contents of the torus world. Any differences in appearance are simply due to the perspective from which we view the contents. Ellis and Williams maintain that a small universe could be empirically indistinguishable from a plane world. For what might seem to be an infinite plane might be instead an infinite set of images of a finite domain. Homogeneity and isotropy may now be explained in a straightforward way. We need no extravagant hypotheses about how the universe has evolved from its initial conditions. As Ellis and Williams say: 'the universe looks homogeneous and isotropic because we are seeing the same region over and over again' at different stages of its history; and 'this is the simplest reason for apparent homogeneity one can imagine!'[33]

So, in such a universe, the Cosmological Principle is trivially true.[34]

In this and the previous chapter, we have examined three central problems connected with the structure of space and time: what standard of simultaneity may we adopt; is the metric of space and time Euclidean; and just how much can we say about the topology of the actual universe? We now have a sufficiently secure basis to clarify the philosophical problems associated with conventionalism about space and time with rather more confidence.

THE UNDERDETERMINATION OF THEORY BY DATA

The claim that theories may be underdetermined by the evidence and so have a conventional character has its origins in the work of Pierre Duhem as well as of Henri Poincaré.[35] Writing in the early years of this century, Duhem argues that there can never be any 'crucial experiment' which allows us to make a definitive choice between two theories. For no theory stands on its own: it is always supplemented by auxiliary hypotheses which are not part of the theory itself. Such hypotheses set down the initial and boundary conditions which can provide the essential empirical background for a theory and its laws. If a theory is challenged by some adverse experimental result, then we may always blame the auxiliary hypotheses rather than condemn the theory itself. So an appropriate adjustment to our auxiliary hypotheses can always save the theory. Many recent writers have found Duhem's message both convincing and attractive. Indeed, the logician Quine supports (and goes beyond) Duhem's position by saying that: 'Any statement can be held true come what may, if we make drastic enough adjustments elsewhere in the system (of beliefs) ... Conversely, by the same token, no statement is immune to revision' (Quine 1980: 43).[36] These ideas have been enshrined within the 'Duhem–Quine thesis' which may be stated provisionally as follows: because scientific theories are underdetermined by the empirical evidence we may always find an incompatible alternative to any given theory. But, we may ask, at what cost?

When observations of the orbit of the planet Mercury were shown to be at odds with the predictions of Newtonian gravitation, scientists looked for some problem in the auxiliary hypotheses. After all, such a tactic had produced tremendous results for the astronomers Le Verrier and Adams when a similar anomaly in the orbit of Uranus led to their joint discovery of Neptune: the hypothesis that the planets stopped at Uranus was at fault and Newtonian gravity was saved – for the theory was able to show how Neptune's presence affects the orbit of Uranus. So it was not entirely surprising when, in the 1850s, Le Verrier postulated the existence of yet another planet, called Vulcan, inside Mercury's orbit – yet another adjustment to the auxiliary hypotheses to keep Newton's head above water. But this time no new planet could be found. Even so, other reasons could be given for the failure to discover Vulcan: for example, any planet so close to the Sun would be obscured by the glare of the Sun, making a definite observation extremely difficult to obtain. It does indeed seem that we might always find some reason to hold on to our theories, however damning the evidence against them might appear to be. Of course, such a reason may lack conviction, the resultant combination of theory and supporting hypotheses might be complex, and the change in auxiliary hypotheses may quite justifiably be called *ad hoc*. As Karl Popper and Richard Feynman both observe, to introduce some hypothesis which is designed to do no more than save a theory in the teeth of the evidence can show a callous disregard for scientific integrity: if we design a crucial experiment, the fiercest test which we can devise for some theoretical position, and if the results of that experiment run contrary to the theory's predictions, then it is better to accept that the theory has been falsified than to invent some *ad hoc* hypothesis to aid our continued belief in the theory.[37]

Does conventionalism have any positive message? Or is it merely a negative response to the scientific realist's belief that scientific theories can and do represent the world in a straightforward and independent way? Both Quine and Reichenbach emphasise the idea of empirical equivalence when trying to clarify the supposed conventional character of our scientific beliefs.[38] But this idea is not altogether clear itself. If we are to get a clear picture of conventionalism, then we need to know just when we may say that two alternative theories are empirically equivalent. For, without some such equivalence between alternatives, we will be able to cite some matter of fact which is likely to offer more support to just one of the two theories. And we also need to be confident that the presuppositions involved in any claim that two theories are empirically equivalent are generally sound.

However, it is difficult to state any convincing thesis of empirical equivalence; most candidates either lack force and are too glib to be useful or, if forceful, make too many problematic or unpalatable assumptions. Consider, for example, this statement:

A T_1 is empirically equivalent to T_2 if both theories are acceptable on the basis of the same available empirical evidence E.

This statement captures the idea that two equivalent theories should rely on the *same* observational evidence, but the two theories might still be empirically distinct. They might make quite different predictions which we have yet to test empirically. So, although the theories may be acceptable on the basis of the same evidence, there remains the possibility that some additional evidence will point unequivocally to just one of the two theories. Also the statement involves no sense of the two theories clashing – so that belief in one theory somehow rules out belief in the other. The idea of a clash is certainly involved in Poincaré's views. Nor is there any indication of the idea that we can always find some alternative to any given theory, as seems to be suggested by the provisional statement of the Duhem–Quine thesis. So we might be more inclined to accept a rather more detailed version of the thesis:

B For any given theory T_1 which is acceptable on the basis of empirical evidence E, there is at least one equally acceptable but incompatible theory T_2 which is empirically equivalent to T_1 and which makes the same empirical predictions as T_1.[39]

Although this seems to be more informative and helpful, three major questions need to be settled before we should embrace B or any similar thesis:

1 What is meant by the idea of empirical acceptability? Generally, a theory may be said to be acceptable:
   a if it is supported to a reasonable degree by the available evidence; and
   b if it is not as yet falsified by any such evidence.

However, what we are to count as a reasonable degree of support does need further clarification – and this should be done within the context of a sound theory of confirmation. Yet the development of such a theory is a substantial problem – for the conventionalist just as much as for the scientific realist.

2 What precisely are the conditions under which two theories are

incompatible? If the idea of incompatibility is to have any force, then it must threaten the position adopted by the scientific realist who believes that what a theory says is at least approximately true of the world itself. Hence, two theories are incompatible when they imply the existence of two radically distinct real worlds – so distinct, in fact, that the realist could not claim that they are *both* approximations to the truth at the theoretical level. This implication carries with it the assumption that a theory is *not* reducible to its observational consequences. For, if it were so reducible, there would be no radical difference between the theories.[40]

3 With respect to what kind of empirical evidence can we say that two theories are equivalent? There are three main categories of evidence which might be cited here:
  a the empirical evidence which is currently available;
  b some restricted domain of evidence, for example: evidence within the context of a given domain such as kinematics (ie excluding dynamics);
  c all possible empirical evidence.

Clearly, only the last category might cause a significant problem for the realist. For the first two will always leave open the possibility that we could decide at least in principle between two alternative views on empirical grounds.

These questions demand some changes in the thesis if the conventionalist is to maintain a clear and forceful position which contrasts sharply with the realist point of view. In order to express such a position, we might move to a stronger version of the thesis:

C For any given acceptable theory T_1, there will always be an alternative theory T_2 such that:
  i both theories are equally well confirmed by all available evidence;
  ii neither theory is falsified by available evidence;
  iii no possible empirical evidence could support only one of the theories;
  iv the two theories imply the existence of radically distinct real worlds.

Points i and ii help the conventionalist to clarify the idea of empirical equivalence. Point iii extends the idea of equivalence across all physical possibilities. And points iii and iv carry the conventionalist attack on realism. Without iii, the realist could say that any equivalence based on available evidence is merely superficial, and that further evidence could, at least in principle, demonstrate that the alternatives are not genuinely equivalent. Without iv, the realist could argue that two empirically equivalent alternatives are both approximations to the same real world and that any differences between the theories are trivial or pose no threat to the belief that over time scientific beliefs are converging on a truthful account of the world.

So, although the conventionalist position does seem to be an attempt to block some of the natural instincts of scientific realists, it is wrong to dismiss it as a merely negative strategy. For, if thesis C is correct, then we have achieved a powerful and positive articulation of the limits of realism within the context of empirical belief. The realist may, of course, challenge C by questioning the implicit assumption that we may isolate an empirical domain which might then serve as a reference point between distinct theories. But, even if we accept C, there are two main options before us:

1 We might adopt C with a thoroughly conventionalist spirit: any choice between alternatives would then be simply a matter of convention – a given theory would be adopted freely as true by convention.
2 We might argue that non-empirical considerations nevertheless allow us to select just one of the alternatives, citing, for example, greater simplicity as the reason for our choice. If such a selection is made on pragmatic grounds, then we might still maintain an essentially conventionalist position.[41] But if there is any suggestion that the reason for our choice has some a priori foundation – perhaps an a priori belief that simplicity is a sign of the truth – then we would not only be parting company with conventionalism, we would be in the invidious position of needing a coherent argument to support such a step.[42]

There is an additional possibility which involves a further modification of the idea of conventionalism but which drops the idea of underdetermination:

D For any given acceptable theory T_1 there will always be an alternative theory T_2 such that:
   i both theories are equally well confirmed by all available evidence;

ii neither theory is falsified by available evidence;
iii no possible empirical evidence could support only one of the theories;
iv the theories are not genuinely distinct since each is reducible to exactly the same empirical basis given that the theories imply no more than their observational consequences.

Point iv here is suggested by Reichenbach's position on geometry noted above. Empirical equivalence is central. But Reichenbach asserts that apparently different geometrical theories are merely expressions of the same theory. This seems to imply that a theory may be reduced to its observational consequences, since Reichenbach argues that two different theories amount to essentially the same 'theoretical' view. He may only say this if any differences at the theoretical level are eradicated during a reduction to the observational level. However, if we accept Reichenbach's argument, we may no longer say that a theory is underdetermined by its observational content. For a theory is no more than its observational content!

Does thesis D involve a commitment to a firm, absolute distinction between theory and observation? Positivists found to their cost how difficult it is to maintain any clear-cut distinction between theory and observation. As Hanson, Kuhn, and others argue, our observational claims about the world seem to depend to some extent upon our wider theoretical perspectives.[43] So we cannot just simply assume that our observational statements are free from theoretical presuppositions – we need to provide convincing arguments for this.[44] However, the above thesis need not involve any disastrous assumptions about the status of our empirical beliefs. Like the earlier statements of underdetermination, thesis D does make an implicit distinction between observation and theory. Each thesis refers to the idea of empirical evidence as if this can be identified independently of theoretical presuppositions. But this distinction need not be absolute; Reichenbach need not propose some *absolute* observational domain. Instead, the empirical evidence cited in an assertion of equivalence between any two theories may be *relative* merely to the two theoretical alternatives involved. An observational domain is indeed required if empirical equivalence is to make sense; but many observational reports may be common to both theories, even if these reports are not found elsewhere.[45]

Is Reichenbach a conventionalist? He is in one important sense:

it does not matter to him which of two acceptable alternative geometrical theories are employed to describe the world. Both theories are 'correct' since each theory amounts to no more than its observational consequences, and these consequences are the same – given empirical equivalence. So the choice of non-Euclidean geometry is a conventional decision even if we have good pragmatic reasons for our decision; Reichenbach suggests considerations of descriptive simplicity as the most persuasive criterion of theory choice. But, unlike the conventionalist who adopts thesis C, Reichenbach must provide an explanation of how a reduction from theory to some observational basis is to be achieved. Even if the observational basis is relative to just two alternative theories, we still need to specify how the complex domain of theoretical terms might be reduced in a coherent way to observational language. Those who have tried to justify any such reduction have been plagued by difficulties.[46]

Hence, thesis C appears to be the conventionalist approach which is the most problem-free and which also poses the most acute problems for the realist, given its strong commitment to the idea of underdetermination. How many of the specific 'conventions' discussed in the earlier sections of this and in the previous chapter carry the force demanded by thesis C? Which of them poses a serious threat to the realist? Although we have been able to raise the questions here, it is beyond the scope of this book to attempt anything more than a preliminary review of these complex issues. We have merely fixed the criteria against which candidates for conventions may be judged.

However, two related general points about conventionalism may be made. If we revise one part of a theory at a theoretical level in the attempt to construct an 'equivalent' alternative to the original, then revisions must be made elsewhere in its theoretical context. If we change an important enough part of the theory, then the subsequent revisions may be so radical that the entire context is fundamentally changed. The two 'equivalent' theories would be radically distinct in almost every way. But we must remember that our scientific theories are embedded not only in a generally accepted scientific context but also in our wider system of beliefs. We simply cannot judge between two alternative theories in isolation from the rest of our beliefs. Consequently, it might be foolhardy for anyone to guarantee that such a radical shift *could not* in principle introduce an empirical difference between two alternative theories with respect to our overall system of beliefs. However, if the two theories differ only in minor ways, so that the fundamental ideas of two 'equivalent' theories are essentially the same, then any conventional feature resulting from such a modest change in outlook is unlikely to embarrass the scientific realist, who typically focuses on the status of the central features of a theory.

Secondly, the conventionalist needs to say why empirical evidence alone should be used to help us decide between two competing theories. Consider just one problem for the conventionalist. Malament's comparison of standard and non-standard accounts of simultaneity reveals that simultaneity in STR depends upon the conformal geometry of its spacetime. The conformal structure of Minkowski spacetime has rather more than a 'background' theoretical role. It plays a central part in the network of models and theories used to describe the world kinematically (STR) and dynamically (GTR and electromagnetism), and even features in discussions of quantum gravity. As Michael Friedman points out, such theoretical concepts seem to have a 'unifying' and explanatory role in our system of beliefs – they provide key links which underwrite our analyses of related phenomena in different physical domains.[47] If, by moving from a standard view of simultaneity, we lose at least some of these links, then we may have a powerful non-empirical reason to resist the move.

The conventionalist seems to be motivated by a deep scepticism about the status of theory. This scepticism also motivates the relationist, who tries to reduce the ideas of space and time to the more concrete concepts involved in material interactions and relationships. Like conventionalists, relationists typically try to minimise the role of theoretical terms in science. This anti-theoretical strategy is frequently advanced by means of arguments for simplicity. The conventionalists ask us to rest content with the relative simplicity of the empirical content of a theory; they see no need to grant high status to theory as such, given their claims about the underdetermination of theory by data. And the relationists frequently wield Occam's razor, trying to cut 'superfluous' entities like space and time from our scientific vocabulary.[48] Both tempt us with a 'simpler' picture of the world in which theory plays at best a secondary role. Although there is clear evidence that principles of simplicity do constrain and structure the development of scientific theories, we should not accept too readily the proposition that ideas of simplicity alone should govern the way we do science. In the following three chapters, we shall explore the relationist strategy and assess the arguments for and against relationism.

5

NEWTON AND THE REALITY OF SPACE AND TIME

INTRODUCTION

A few pleasantries in a handful of letters were all that passed between the two giants of science and mathematics in the seventeenth century: Isaac Newton and Gottfried Leibniz. In these letters there is little or no suggestion of the gulf between them on their attitudes towards the reality of space, time, and motion. Only in Leibniz's correspondence with Samuel Clarke and Christian Huygens do we see any detailed contemporary debate on the issues which divide Newton and Leibniz.[1] Much of this debate is concerned with theological issues turning on the nature and the powers of God, but throughout there is a strong desire to clarify the concepts of space and time. The debate is provoked by Newton's *Principles of Natural Philosophy* (*Principia*) and particularly by his assertion that space and time are absolute entities.

Newton's *Principia*, published in 1687, presents us with a powerful and persuasive analysis of space, time, and motion. Space is, in Newton's view, essentially an absolute, independent, infinite, three-dimensional, eternally fixed, uniform 'container' into which God 'placed' the material universe at the moment of creation.[2] Time is an absolute, independent, infinite, one-dimensional, fixed, uniform 'framework'. Absolute (as opposed to merely relative) motion is motion through space itself – an object which is really moving changes its absolute position in space continuously. It is impossible for us to say whether any object is at absolute rest in space; for there is no experiment or observation which will allow us to single out any one frame of reference as absolutely at rest. Newton's laws of motion imply an equivalence between all non-accelerated frames of reference. This equivalence is the essential foundation of the classical 'Galilean' principle of relativity, which states that the laws of motion apply in the same way to all non-accelerated systems. So we cannot use the laws to help us decide which if any of two or more bodies in relative uniform motion is really in motion or really at rest. The 'fixed' stars might seem to be fixed and at rest in space – but, for all we know, the entire material universe might be moving at a uniform velocity through space. But we may say of certain objects that they really are moving. Objects which are accelerated experience 'inertial' forces – the kind of forces at work when we are forced back into our seats as a car or plane accelerates. If I am enclosed in a box and experience no inertial forces, then I can do nothing to say whether the box is at rest or moving at a constant velocity relative to any background framework I choose – including space itself. But, as soon as I experience inertial forces, then I may be certain that the box is accelerating in some way; it may be accelerating in a straight line or it may be rotating or both. This empirical fact provides the foundation for Newton's 'proof' of the existence of absolute space: a proof which may be easily extended to provide an argument for absolute time as well.

Leibniz resists Newton's claims about space and time strongly. In his correspondence with Samuel Clarke, Leibniz defends the adoption of relational concepts of space and time. Although his defence turns on the acceptance of a rationalistic philosophical system, Leibniz nevertheless presents some formidable challenges to Newton's absolutist perspective. In this chapter we will explore the issues which divide Newton and Leibniz. And this debate will turn out to be of more than merely historical interest, for the problems discussed remain central to our understanding of the nature of space and time as described in modern physics.

ABSOLUTE SPACE AND TIME

Newton, in his celebrated rotating bucket and two globes thought experiments, asks us to consider the general properties of rotating systems; see Figure 25 (p. 101). We may reconstruct his argument in the Scholium on space and time in the *Principia* along the following lines:[3]

1 Our general experience leads us to link occurrences of inertial forces with accelerations such as rotations. So we might argue

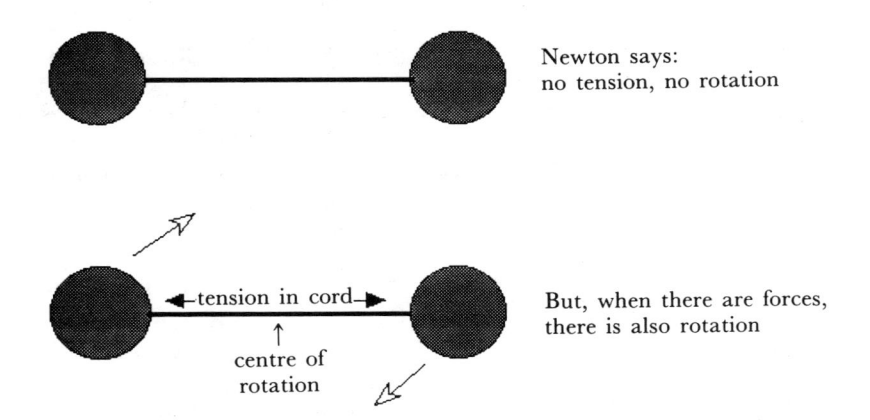

Figure 25 Newton's thought experiment: rotating spheres

(Top) Newton says: no tension, no rotation

(Bottom) But, when there are forces, there is also rotation

← tension in cord →
↑ centre of rotation

that any rotating system experiences inertial forces as a consequence of its rotation; for example, a system of two globes connected by a cord rotating about a common centre would experience a tension force along the cord.

2 With a system like the two globes and cord there are two possible situations: either there is a tension in the cord or there is no tension.

3 In both cases, the relative positions of the globes and cord are always the same.

4 So, if we restrict our attention to the system itself, then the only way we might tell that it is rotating (or accelerating in some other way) is by checking for tension in the cord.

5 We might suppose that we can always check for rotation by looking for relative motion between the system and some background frame of reference like the fixed stars.

6 But we can easily imagine the system in an otherwise empty space – in an immense void, as Newton calls it.

7 In this case, we are unable to rely on a material background frame of reference. But we can still be sure whether or not the system is accelerating by checking for signs of tension in the cord.

8 If there is tension, then we can justifiably say that the system is accelerating 'absolutely' with respect to space itself.

9 And in such a case the source of the inertial forces must lie in

some 'inertial' interaction between the accelerating system and space.
10 Therefore, we cannot explain the presence of inertial forces without an essential reference to space itself. In this sense, space may be said to be absolute – it is an irreducible element in our physical description of matter and forces.

Newton's argument for absolute space may be readily extended to give us a 'Newtonian' argument for absolute time.[4] Inertial forces in the globe system indicate: that there really is a rotation; and, therefore, that the velocity of each globe is continuously changing, because velocity depends on direction as well as speed. The changes in direction and therefore in velocity are changes in time. But in an otherwise empty space there is no changing material framework to which this change may be referred. So the change is relative to a non-material temporal structure: namely, absolute time.

Steps 6 and 7 in this argument are crucial. They carry with them the assumption that the results of dynamical experiments would be no different even if there were no other matter in the universe. If we allow this assumption, then it is hard to see how we might rule out space as an irreducible entity in its own right. It might seem that Newton is going too far beyond the evidence in making this assumption; but he is simply relying on a strategy common amongst physicists. Because of the apparent universality of the effects under discussion, Newton sees no reason why he should not abstract from actual conditions to more general cases. Indeed, he has every reason to do so. Laws of nature are usually said to hold in counterfactual as well as factual situations. A factual statement describes a situation which has happened or is happening. A counterfactual statement describes a situation which might have happened or might be happening, but which in fact has not taken place. Indeed, we might reasonably argue that it is the provision of support for counterfactuals which distinguishes laws from accidentally true generalisations.[5] This process of abstraction involving the application of laws to counterfactual situations has met with tremendous success throughout the scientific enterprise, given that it allows us to predict how things will turn out in unfamiliar circumstances. Ernst Mach, writing in *The Science of Mechanics* in 1883, provides perhaps the strongest challenge to the assumptions behind steps 6 and 7; we shall review his argument in the next chapter.

We should also note the assumption behind step 8. Newton regards motion as a relation between two objects: when one of these 'objects' is space itself the motion is absolute. But why should we accept that motion is a relational property at all? Lawrence Sklar suggests that we may treat motion as a brute fact about an object rather than a relation between two or more objects. We shall consider this interesting challenge to one of Newton's fundamental assumptions in the final section of this chapter: 'Absolute motion without absolute space?'

We might be inclined to add the claim that space is a substance consisting in 'a substratum of space points or regions that underlie bodies' to Newton's argument.[6] We shall see later in this chapter that Samuel Clarke does precisely this in his defence of Newton. There is some justification for Clarke's move. Newton talks freely in the Scholium on space and time in the *Principia* of 'parts of space'; and he does seem to treat space as a substance of some sort in its own right. But the argument in the Scholium by itself fails to provide any basis for this further claim about substantival *points*. It is an argument for absolute motion and for space as an irreducible element in our description of motion. It is not an argument for substantivalism, if we mean by that the claim that space consists (of probably a continuum) of substantival points. If we accept Newton's argument, then we may have to concede that space is a substance of some sort. But we need further justification for the move to treat space as a substratum of points. Of course, the success of Newton's geometrical description of motion may be cited in favour of the claim about points. For this approach to the problems of motion does rely on the idea of a continuum of points. However, as we saw in Chapter 1, we have some reason to be suspicious of the status of the continuum. So we may be inclined to reserve judgement on the claim that confidence in the existence of space as an essential element in our description of motion should automatically incline us towards the view that space is a substratum of points. We shall discuss this issue further in Chapters 7 and 10.

MATTER IN THE NEWTONIAN WORLD

Newton's conception of the physical world is essentially mechanical and corpuscular. Material bodies which consist of tiny corpuscles interact with one another in a vast spatial container according to the mechanical rules set down in the *Principia*. In contrast to Descartes's view of 'contact' material interactions which dominated the period, Newton's account of gravity allows matter to influence matter at a distance.[7] His laws of motion and gravity are said to hold, not just for our own limited region of space – the solar system – but everywhere and for all times. And these laws govern the behaviour of corpuscles or particles which are 'solid, massy, hard, impenetrable, moveable'.[8] Newton's views on matter, where articulated, are generally close to those of Robert Boyle and John Locke.

John Locke, who had followed with keen interest and who had even assisted the experimental work of Robert Boyle in Boyle's Oxford laboratory, helped to consolidate the late seventeenth-century notion of matter in his *Essay on Human Understanding* published in 1690. Boyle's own publications in the 1660s, including *The Origin of Forms and Qualities*, had a powerful impact upon Locke's thinking. Locke accepted Boyle's materialist and mechanical explanations of a physical world populated with corpuscles or particles of matter. Locke, again in accord with Boyle, realised that many of the things that we say about matter are suspect in the sense that they may not be true of matter itself. They may merely be comments on the ways we perceive material things. So Locke adopts Boyle's distinction between the primary and the secondary qualities of matter: between the primary properties genuinely possessed by matter and the secondary properties which arise only in our perceptions of material objects.

Locke, who corresponded regularly with Newton, was fully in sympathy with Newton's empirical characterisation of the material world. He shows us just what our epistemological commitments are when we accept the idea of matter assumed in the *Principia*. When we think of any material object, we must think of it in terms of its primary qualities: as extended; as in some state of motion or rest; as having a surface and an interior; and as being either a single entity or some collection of entities. But when we think of an object as having taste, for example, there is nothing in the object *per se* which impels us to regard such a quality as inherent in the object. A sweet taste to one person may be undetectable by someone else. An object may be blue at one moment, but, when the light is switched off, no colour is there at all. We may readily think of matter without conceiving it as coloured or tasty. However, Locke believes that it is impossible for us to conceive of matter without thinking of it in terms of its primary qualities. If we take a grain of wheat and subdivide it again and again, Locke tells us that it will still possess the same primary qualities. Primary qualities are therefore the most general and basic characteristics of matter. Each quality is a basic building block of the Newtonian conception of matter; and Newton's material universe may be satisfactorily characterised in terms of these essential conceptual building blocks.[9]

Although Locke and Newton seem to be in agreement on the basic idea of matter, Locke's earlier thinking on space and time is at odds with Newton's views. In his Journal for July 1676, Locke suggests that space and time may be relations amongst material things and events and not substantival physical entities:

> And as we can in our imagination apply that measure of time, which is but the motion of some body, to duration, where there is neither body nor motion, so we can apply that measure of extension, which is only in bodies, to space, where there is no body at all, though that duration and that space without the existence of any other thing be in itself really nothing. ... But when things really are, both duration and space, that is considered as so much space between them, are really a relation commensurable to our measures of time and extension.
>
> (Locke 1676)[10]

However, by 1690, Locke speaks quite happily of 'pure space' and of 'duration in itself ... going on in one constant, equal, uniform course' even before the moment of creation.[11] The event before this change in attitude was, of course, the publication of the *Principia*.

LEIBNIZ AND RELATIONISM

Leibniz's metaphysical beliefs tend to obscure his attacks on the idea of absolute space. Leibniz draws a distinction between the 'phenomenal' world of appearances and the world as it really is: our senses give us access to appearances, but only reason allows us to throw back the veil of perception. Although our senses seem to tell us that the world consists of material objects occupying space and persisting through time, he says that we have no reason to trust our perceptions. Instead reason should be our guide to the structure of reality. Leibniz tells us that we may understand how things actually are only by deduction from undeniable principles. Like many philosophers before and after him, he regards the deductive methods of mathematics and geometry as an attractive model for philosophical speculation. Using such clearly 'rational' methods he constructs a mysterious world of monads – simple metaphysical substances. This world is neither the spatio-temporal nor the material realm which sense experience suggests. It is a world to which rationality can give us access; but this access is limited by our finite natures – only God has the infinite capacities needed for a complete grasp of reality.

In dealing with the problems of space and time, Leibniz moves somewhat uneasily between the realm of monads and the physical world. The monadic realm is sometimes characterised as 'spatial' and 'temporal', but not in the senses we usually employ for these terms. Leibniz seems to claim that what we see as spatial and temporal properties are merely analogous to the underlying structures of reality; hence that the physical world seems to have spatio-temporal properties is a consequence of our imperfect perceptions of the structures of the monadic realm. Leibniz's denial of any independent 'absolute' status for space and time in the physical world derives from his characterisations of the monadic realm. The relations between monads underwrite the relations which we see as holding between the objects of the phenomenal world. Leibniz compares these relations to those which hold between the members of a family tree: no given relation has any fundamental status for any member of the tree, for 'he who was a father, or a grandfather, might become a son, or a grandson'.[12] But since the monadic realm is, perhaps necessarily, opaque, he concentrates his attacks on the notions of 'absolute' space and time within the context of the phenomenal world. So, when he says that space and time are 'merely relative' since space is 'an order of things' which coexist and time is 'an order of successions' of things, Leibniz is not thinking of 'things' as his metaphysical monads but as the regular material objects of the phenomenal world.[13]

Leibniz's argument against absolute space has its clearest expression in his correspondence with Samuel Clarke during the period 1715–16. The correspondence consists in five letters from Leibniz and five replies from Clarke, and a spirited debate was only brought to an end by Leibniz's death. Much of the correspondence concerns the relationship between God and the world, but their respective beliefs about the nature of space, time, and motion emerge reasonably clearly and the essential differences between them do not rely on any particular view of God. However, their debate is plagued by numerous misunderstandings, mostly on the part of Clarke; and it also suffers from Leibniz's ultimate failure to provide an adequate answer to Newton's argument for absolute space as an essential element in our description of motion. But, as we shall see in the next two sections, a natural extension of Leibniz's ideas does suggest a possible answer: absolute motion without absolute space or time.

Clarke had worked closely with Newton and the correspondence gave him a golden opportunity to defend the Newtonian worldview. Leibniz argues that the Newtonian case for 'absolute space and time' involves contradictory beliefs and must therefore be rejected. In his third letter to Clarke, Leibniz employs two important principles, the Principle of Sufficient Reason (PSR) and the Principle of Identity of Indiscernibles (PII), as axioms of his arguments against the Newtonian view of space and time:

PSR 'nothing ever happens without there being a cause or at least a determining reason for it';[14]

and, as a consequence of PSR,

PII 'there are not in nature two real, absolute beings, indiscernible from each other [by God]; because if there were God and nature would act without reason, in ordering the one otherwise than the other' and so two indiscernible things are identical – i.e. the two are in fact one and the same thing.[15]

Leibniz adds two 'plausible' assumptions (3 and 4 below) to these principles, and then tries to show that the Newtonian position is incoherent:

1 A rational God must have a sufficient rational reason to act or to make a choice (given PSR).
2 If A is indiscernible from B, then A is identical with B (given PII).
3 Space is an infinite, non-material 'absolute' container for matter.
4 If space exists as an independent container, then matter may be placed in that space howsoever God (who is omnipotent) chooses.
5 God might choose to place matter in space in a given position

and orientation, thus creating a universe X with a given configuration relative to space.
6 He could have chosen to place matter in a different position and orientation, thus creating a universe Y such that the relative positions of the material contents to one another are just as in X, but in which these contents are placed differently in space, e.g. the contents of Y may be rotated through 180 degrees compared with those of X.
7 Space is absolutely uniform, i.e. all its constituent parts are alike; hence the matter distributions in both X and Y will be related to the background spatial container in the same way.
8 Given 7, there is no discernible difference between the two possible universes X and Y.
9 Therefore X and Y are indistinguishable.
10 Given 2 and 9, X and Y are the same thing.
11 A rational choice cannot be made between two objects which are in fact 'the same thing' since there is no rational choice to be made.
12 Given 1, 10 and 11, a rational God does not have a sufficient reason to choose between X and Y.
13 Therefore God has no choice in how matter is placed in space.
14 Given 4 and 13 (the denial of the consequent of 4), the antecedent of 4 is also false (by *modus tollens*).[16]
15 But assumption 3 is the antecedent of 4: hence assumption 3 (space is an independent 'absolute' container for material things) must also be false.

Leibniz applies a similar argument against 'absolute' time. Instead of asking whether or not God has a free choice as to how matter is placed, Leibniz asks if God may choose the moment of creation. He then says that there could be no rational choice between a universe in which matter is created at a given time and one in which matter is created a year earlier. And, just as in the case of space, there seems no discernible distinction between the two possible universes – their histories would be identical. So we must also reject the idea of time as an independent 'absolute' background framework for change in the material world.[17]

CLARKE'S DEFENCE OF NEWTON

Samuel Clarke responds to Leibniz's arguments by challenging the status and power of PII. He says:

> Intelligent beings are agents . . . they have active powers and do move themselves, sometimes upon the view of strong motives, sometimes upon weak ones, and sometimes where things are absolutely indifferent. In the latter case, there may be very good reason to act, though two or more ways of acting may be absolutely indifferent; [and]
>
> Two things, by being exactly alike, do not cease to be two. The parts of time, are as exactly like to each other, as those of space: yet two points of time are not the same point of time, nor are they two names of the same point of time.
>
> (Clarke 1716)[18]

So, although there may be no apparent difference between the universes X and Y in Leibniz's argument, Clarke claims that they are nevertheless distinct – all points of space and time are indeed alike but they are certainly not the same points. Hence PII does not hold: for two indistinguishable things are not one and the same thing at all. But he does not deny PSR since he accepts that God will always have some reason to act or choose. Like the Stoics in the years after Aristotle, Clarke sees no problem in imagining the entire material world being moved in an infinite space.[19] However, he questions the link between PSR and PII when he says that a choice will be possible even in cases where two objects are exactly alike.

Clarke's defence of the Newtonian position relies on the statement that, despite the uniformity of space and time, they nevertheless have distinctive points. This leads to the claim that space and time are quantities in themselves rather than mere relations between objects. The main argument for this claim appears in Clarke's fourth reply to Leibniz and depends upon Newton's distinction between real and apparent motion, used to good effect in the rotating bucket and two globes thought experiments.[20] This argument, like Newton's, does not depend upon the powers of God and his interactions with the physical world. It turns on an empirical problem which Leibniz must confront if he is to vindicate his relational viewpoint. The following is a reconstruction of Clarke's argument:

1 If the points or regions of space and time can be shown to have distinct characteristics which do not themselves depend on matter and the relations between material objects, then space and time have an independent reality as quantities in their own right.

2 If and only if an object occupies distinct sets of points or distinct regions at distinct times, then it is really moving through space and in time.
3 If and only if an object is really moving through space and in time, then it experiences inertial forces when its state of motion changes.
4 Objects do experience inertial forces when their states of motion are changed.
5 Given 3 and 4, objects really do move through space and in time.
6 Given 2 and 5, objects occupy distinct places or regions in space at distinct moments of time.
7 Hence, the points or regions of space and time have distinct characteristics.
8 Given 1 and 7, space and time have independent existences.

Assumption 2 forges a link between the reality of points or regions in space and absolute motion. We shall see in the final section of this chapter that precisely this link is challenged by Sklar's argument that absolute motion may be a brute fact about an object rather than an assertion of a relation. So we should not leap from 'the object is moving' to 'the object is moving with respect to some other object'. This clearly requires some fundamental changes in attitude towards motion. If we were to make such changes, then neither Clarke's nor Leibniz's *specific* approaches to the problem in hand would be appropriate, for, as we shall see, the terms of the debate would be changed dramatically.[21]

Assumption 3 above relies on Newton's analysis on real and apparent motion and as we have seen the problem of inertial forces is most acute in the context of rotation. The difference between (a) a large wheel turning from an observer's point of view and (b) the observer appearing to move from the point of view of someone attached to the outer rim of the wheel may be traced to the fact that only the wheel and its 'rider' experience inertial forces. That is why we may say the wheel is really moving and the observer is not.[22]

Leibniz had already considered the problems of inertial forces in detail during a correspondence with the physicist Christian Huygens during 1695–5.[23] Their letters indicate that they had reached a measure of agreement concerning the relational nature of motion, although Huygens allies himself in spirit and word with much of the Newtonian view. Each claims to have resolved the problem of inertial forces, but only what seems to be Huygens' purported solution has been discovered. In this, Huygens shows that he appreciates the need for an account of the origin of inertial forces, but simply restates the problem in terms of velocity differences. He argues that any forces which are experienced by an object attached to a rotating object are due to relative motion alone: the fact that an object attached to the circumference of a spinning wheel experiences inertial forces is explained by the relative motion between this object and the rest of the wheel. Although the object and some parts of the wheel may be moving to the left at a given instant, other parts of the wheel will be moving to the right. So a spinning wheel experiences inertial forces not because of its rotation relative to any absolute framework, but because of the relative motions of the various parts of the wheel. Huygens' solution does have the virtue of recognising the objective significance of differences in velocities: such differences may obtain even when the relative positions of bodies or of parts of bodies remain the same, as in the spinning wheel example. Two distinct points on the circumference of the wheel may have the same 'speed' but they will have different velocities because they are moving at any given instant in different directions. But Huygens fails to grasp the problem of inertial forces completely. For we may easily transform away the relative motions by viewing the spinning wheel from a frame of reference which is rotating along with the wheel. There is no relative motion and, as far as we can *see*, there are no velocity differences. But there are plenty of forces! This, of course, is exactly the problem raised in Newton's thought experiments. Unless Leibniz tackles this fundamental problem, we can hardly regard him as victorious in his dispute with Clarke.

Clarke directs Leibniz to Newton's analysis in the *Principia*, which takes us right back to the heart of Newton's argument for space as an 'absolute' independent element of reality. And, as we have seen, a similar argument may be constructed for absolute time. Clarke's argument rests on the same premise as Newton's: inertial forces provide a clear indication of 'absolute' motion through space and in time. Clarke is rather vague about the idea of space and time as absolute 'quantities'; but it seems reasonable to suppose that he takes the natural inference of Newton's argument to be that space and time have essentially the same ontological status as matter itself. They are not qualities of other things; but they are a fundamental (if non-material) quantity.[24] Just as a material body can have real effects on other bodies, so space and time have real effects on objects. Hence, space, time, and matter are all in some sense real quantities.[25]

In his fifth and final letter, Leibniz says that he can find nothing in Newton's discussion in the *Principia* 'that proves, or can prove, the reality of space in itself'.[26] This is because he refuses to accept any link between inertial forces and motion in space. He grants that the presence of inertial forces in a body indicates that the body is in motion, but this motion is merely relative to other objects rather than to space itself. The cause of inertial forces need not be traced back to motion with respect to space and time; for when a body is really in motion 'the immediate cause of the change is in the body' itself.[27] Clarke dismisses this response as inadequate, and refers the reader back to Newton's argument. In the absence of Leibniz's missing 'solution', what is needed is a straightforward answer to the question: how may we explain the presence of inertial forces in some bodies and their absence in others without invoking space and time themselves? The clue to a possible answer lies in Leibniz's rather obscure comment above which suggests that the source of inertial forces lies within bodies themselves – undoubtedly a reference to his notion of *vis viva*, an active internal force, the possession of which is an indication of true motion.[28] This suggests the possibility of true 'absolute' motion *without* the need for absolute space as a global reference background for motion – and, as promised, this possibility will be considered in the next section.

Leibniz offers us a physical world in which only material things have any metaphysical basis. His programme is reductionist in the sense that all spatial and temporal relations may be reduced at the physical level to relations between material bodies (and at the metaphysical level to relations between monads).[29] In a sense, Leibniz proposes a simple monist picture of the world, even if this monism is only fully effective at the monadic level where the monads themselves are characterised as pure spirit. But Newton presents us with a more complex physical world: not only must we accept material things as fundamental, but space and time too are irreducible non-material 'substances'. It is tempting to regard Newton's physical world-view as essentially dualist, and many writers have done just this.[30] But Newton gives us no reason at all to suppose that he thinks that space and time are the same kinds of thing. His physical universe has three distinct kinds of 'substance':

112

matter, space, and time. And at the metaphysical level, spirit is added to these three.[31]

Although the problems of Newtonian space and time are most frequently set in the context of the Leibniz-Clarke debate, there is a good deal of evidence to suggest that Newton's conception of the physical world arises at least in part from his opposition to Descartes's physical theories.[32] Indeed, the ideas of Boyle on matter and its properties, the influence of which may be seen in both Locke and Newton's thinking, are in part a clear, critical reaction to the Cartesian view. Descartes argues for a reductionist point of view, but, unlike Leibniz, he takes spatial extension to be not just the most fundamental aspect, but also the distinguishing feature of matter; hence matter may be reduced to spatial concepts. For a brief time, until Newton's views gained wide acceptance, Descartes's position remained popular.

ABSOLUTE MOTION WITHOUT ABSOLUTE SPACE?

Two rockets in space, far from any star, are alongside each other. One rocket carries Peter; and Paula is in the other. They seem to be quite weightless, floating in a force-free environment. The classical principle of relativity implies that there is no way for them to discover whether they are at rest with respect to space itself, or whether they are rushing along at an enormous but uniform velocity. Then Peter looks out at Paula's rocket and sees it accelerate rapidly away. And Paula looks out and sees Peter's craft accelerate rapidly away. There is no kinematic difference between Peter's perspective and Paula's: both astronauts measure the same relative acceleration between the two rockets. But there may be a dynamical difference. Peter may feel an inertial force whilst Paula remains force-free, or vice versa. If Paula experiences such a force, then she may say that she is accelerating absolutely. But is this acceleration a brute fact about Paula and her rocket? Newton and Clarke do not even question their basic assumption that motion is always relative to some 'thing' – whether this is a material object or space itself. However, Sklar suggests that we might say that Paula simply feels inertial forces and that any object which experiences such forces is 'in motion'.[33] This suggestion seems a natural extension of Leibniz's rather undeveloped view of inertial forces as arising from the internal characteristics of material objects and not from some dynamic relation to an external frame of reference.

Newton explains the presence of inertial forces by invoking absolute motion (i.e. motion relative to space itself). Newton's account suggests that there is some kind of causal interaction between space and matter: space acts on an accelerating object in such a way as to produce inertial forces. It is interesting to note that, contrary to any expectation that Newton's third law of motion might give us, there is no reaction of matter on space, since Newton tells us that space is not influenced in any way by its material contents.[34]

Sklar does not seek an external explanation of the inertial forces: 'real' motion is just something we say is happening to an object when inertial forces are experienced. But this 'real' motion is no more relational than the force itself. Just as it is a brute non-relational fact about an object that it has so many atoms and therefore so much mass, so we may think of the experience of a force as a brute fact about an object. Material objects just happen to be divided into two sets:

1 those objects which experience inertial forces; and
2 those which do not experience inertial forces.

Sklar suggests that there may be nothing to be gained by adding the claim that objects in set 1 are *really* in motion *with respect to space*. Newton might respond by saying that what we gain is the powerful concept of an inertial frame of reference. Accelerated motion is always motion relative to all inertial frames, one of which is space itself. But, since the concept is introduced in order to distinguish the members of set 1 from those of 2, we might argue that there is an element of circularity here and that we need some independent reason to suppose that absolute space is required in our dynamical descriptions. However, Sklar's view seems to demand a radical shift in the way we think about motion. So we might be inclined to dismiss the idea of motion as a brute fact simply because it does not capture our basic intuition that motion in general makes sense only when we think of it in terms of some kind of relative movement.

Michael Friedman argues that space in classical physics provides a unifying principle, drawing together the otherwise separate domains of classical mechanics and electromagnetism.[35] All motions may then be referred to exactly the same background framework: space. Similarly, motions in modern physics may be referred to a spacetime framework which unifies relativity and electromagnetism. What makes the idea of absolute space especially powerful in Friedman's view is the fact that it plays the same explanatory role in distinct theories. Even if we accept Friedman's claim that unification brings better scientific explanations, the roles of space and spacetime as unifying frameworks may still be challenged by Mach's argument that the only relevant background arena for motion is the *material* universe. We shall therefore consider this argument in Chapter 6. Although Sklar's suggestion may seem to lack explanatory power and may seem to involve an extremely counterintuitive idea of motion, John Earman claims that the difficulties before standard absolutist and relationist programmes are so great that we should consider the suggestion as possibly the best answer to the problems of the reality of space and time. This claim will be considered in greater detail when we examine Earman's critique of absolutist programmes in the final section of Chapter 7, together with some further discussion at the close of Chapter 10.[36]

6

MACH AND THE MATERIAL WORLD

INTRODUCTION

In his study of motion 'De motu' written in 1721, Bishop George Berkeley dismisses the notions of absolute space and time as being without meaning. Berkeley believes that only sense experience may underwrite meaning; and, since space and time have no foundation in our sense experience, we have no reason to accept them as meaningful words. This line of argument is one pursued nearly two centuries later by Ernst Mach and by others of a positivist persuasion. However, Mach adds his own unique touch to the argument – combining the empiricism of Berkeley with a deep respect for simplicity in science.[1]

Mach's argument is driven by two distinct motivations: the desire, like that of Berkeley, not to go beyond the evidence of direct sense experience; and the wish to achieve the greatest possible simplicity in characterising the physical world. There is no doubt that his dislike of the idea of absolute space stems in part from the fact that space itself seems to be unobservable. Mach says that only the objects of sense experience have any role in science: the task of physics is 'the discovery of the laws of the connection of sensations [perceptions]'; and 'the intuition of space is bound up with the organisation of the senses ... [so that] we are not justified in ascribing spatial properties to things which are not perceived by the senses'.[2] And Mach seems to be more than willing to wield Occam's razor to excise any superfluous entity from our scientific descriptions. His remarks on simplicity in science show that he takes economy to be of central importance; he says that 'the fundamental conception of the nature of science [is the] economy of thought' and science's 'goal is the simplest and most economical abstract expression of the facts'.³

Such assertions form the philosophical background to Mach's attack on hypothetical entities. Atoms as well as absolute space receive a strong rebuff from Mach. His attitude may be summed up thus: if you cannot see or otherwise sense a purported entity and if it is not an essential element of our physical descriptions, then do not grant anything more than instrumental status to the entity – at best any reference to the entity is shorthand for a collection of observational descriptions; at worst references to an entity may carry with them too many metaphysical associations to admit them to our scientific descriptions.

Mach's most important role in the debate about space and time may well be as an *agent provocateur* rather than as an active participant in his own right. Yes, his challenge to Newton's argument is significant. However, the main significance probably lies in how others have been motivated by the challenge. For Mach was unable to fulfil his own dream of a theory of motion without space and time as essential 'absolute' elements. Many, including Einstein, admired Mach's fierce scepticism about the status of theoretical entities, even if few agreed fully with Mach's general philosophy of science. Although some writers claim that Mach ultimately failed in his attacks upon Newton and that he propounded a disastrously flawed philosophy, there is little doubt that Mach's influence has been both far-reaching and long-lasting.[4] Three aspects of this influence are noteworthy:

1 Einstein was deeply impressed by Mach's *Science of Mechanics*, and, for some years, he believed that relativity might be the realisation of one of Mach's aims: a theory of dynamics with no essential reference to space itself.[5]
2 Despite Einstein's admission that his own theory seemed to fall short of this aim, Mach's ideas also inspired eminent physicists throughout the twentieth century to construct variations of relativity which might fulfil that same dream.[6]
3 Mach also stimulated many in the world of philosophy: in particular logical positivists, forming one of the most influential philosophical movements of the mid-twentieth century, at times lent heavily upon Mach's positivism in formulating their philosophical programme.[7]

Even if Mach's scientific dreams are never realised, his role as an inspiration to a generation or more of scientists and philosophers marks him out as one of the most important contributors to our ideas about space and time.

MACH'S RELATIONISM

Passages throughout *The Science of Mechanics*, published in 1883, show Mach to be unsympathetic to Newton's concept of space as an absolute frame of reference for motion. But, unlike Leibniz, he does try to provide a cogent response to Newton's empirical argument for absolute space. He goes right to the heart of Newton's argument and challenges the basic presumption behind steps 6 and 7, which we may recall from the last chapter are:

6 But *we can easily imagine the system in an otherwise empty space* – in an immense void, as Newton calls it.
7 In this case, *we are unable to rely on a material background frame of reference*. But we can still be sure whether or not the system is accelerating by checking for signs of tension in the cord.

Newton has no qualms about assuming that the dynamical effects observed in an Earth-bound laboratory will apply in all cases, possible as well as actual. The fixed stars may act as a convenient global frame of reference for motion, but Newton clearly believes that their presence is merely incidental and therefore they should not have any primary role in the analysis of motion. But Mach objects that we have no way of determining what might happen in circumstances radically different from those observed – where the entire bulk of the universe is removed. So what is the sense in talking about 'empty space' with no material contents? We must deal with the universe in which we live and not build speculative castles in the air. He says that:

> It is scarcely necessary to remark that in the reflections here presented Newton has again acted contrary to his expressed intention only to investigate *actual facts*. No one is competent to predicate things about absolute space and absolute motion; they are pure things of thought, pure mental constructs, that cannot be produced in experience. All our principles of mechanics are, as we have shown in detail, experimental knowledge concerning the relative positions and motions of bodies. . . . No one is warranted in extending these principles beyond the boundaries of experience. In fact, such an

extension is meaningless, as no one possesses the requisite knowledge to make use of it.

(Mach 1883: 280)[8]

Mach does not see why we should accept Newton's belief that the fixed stars are incidental in the analysis of motion. So Newton is accused of ignoring the role of the fixed stars when what is at issue is precisely that role. Our scientific reasoning should be limited to the world as it is, the world of observation and experiment. Hence, we have no reason to trace the origin of inertial forces to anything other than material bodies. To do otherwise is to engage in fruitless and probably meaningless speculation. On a global scale, our reference frame must be 'the entire universe' and we should therefore look for a dynamical theory which accounts for inertial forces in material terms and which therefore does not rely on the probably superfluous idea of absolute space.[9]

In attacking Newton, Mach presents a forceful challenge to the idea that our laws might operate in *any* counterfactual circumstances. He argues in *The Science of Mechanics* that we should refer motions to a material framework – that of the material universe as a whole. He sees no compelling reason why we should rely upon 'metaphysical' artefacts like absolute space when all motions might be referred instead to a physical frame of reference. The inertial forces experienced by accelerating bodies might then be the result of some kind of global interaction with the rest of the matter in the universe. Mach speculates much about the possible material interactions, local as well as global, which might give rise to inertial forces. He asks, for example, whether an enormously thick-sided bucket set in rotation might have immediate and observable local inertial effects on any 'stationary' water within due to the resulting *relative* rotation between bucket and water. However, Mach offers us no detailed positive explanation of how such a material interaction might operate. Mach fails to raise, let alone answer, such important questions as: do inertial forces fall off with distance according to an inverse function or an inverse square function? He simply stops short at his negative critique of Newton's views. Einstein, who had studied *The Science of Mechanics* as a young man, admired Mach for his scepticism but was sceptical himself about the positive side of Mach's ideas. Einstein, perhaps more than anyone at the turn of the century, realised that much work had to be done before science could provide a coherent account of inertial forces without reference to space and time themselves. Mach did not achieve his desire for an economical account of dynamics containing no essential reference to space and time.

Newtonian theory demands such references; and therefore the best that may be said for Mach's argument is that it could point the way either to a new theory or a new variation of Newtonian theory in which matter itself is the source of inertia. But we need more than rhetoric to persuade us to give up an established and successful scientific theory. Two main alternatives face us:

1 We need to be shown that irreducible references to theoretical entities such as space are wrong in principle, so that any theory containing such references may be rejected on those grounds alone. If we take this path, then we shall require a very strong statement against theoretical entities and statements.
2 We need to look for an alternative theory which makes no essential reference to theoretical entities such as space and time, achieving what Mach might regard as a simpler as well as an observational description of the physical world. In this case, we shall still have to justify the implicit *preference* for observational over theoretical statements.

So, if we are inclined to be sympathetic to Mach's approach, we need to explain why we should adopt his empiricist bias. There are two central elements in Mach's approach to the problem of motion: his positivistic outlook and his ideas about simplicity. There is some reason to think that his theory of economy in science both constrained and informed his positivistic outlook. So, before examining this outlook further, we shall explore the main ideas in his theory of economy.

SIMPLICITY AND SCIENCE

Mach's ideas on simplicity began to emerge as early as 1861, some twenty years before the publication of his attack on Newton in *The Science of Mechanics*.[10] In that twenty-year period Mach developed perhaps the first detailed account of the role of economy in science. His admiration for both Galileo and Newton was based in part on the elegant and simple treatment which these two giants had given to their respective ideas of mechanics and motion. And, in their work and in the work of other great scientists, Mach saw three distinct but inter-related kinds of economical attitudes at work.

MACH AND THE MATERIAL WORLD

1 Mach recognised that the scientific enterprise is far too vast and complex to be managed if we are distracted by too great an attention for the fine detail of the physical world. Scientists look to simplifying relationships and formulae so that they have some chance of success in their attempt to characterise the world. Mach readily admits that 'We never reproduce the facts in full, but only that side of them which is important to us, moved to this directly or indirectly by a practical interest. Our reproductions are inevitably abstractions' (Mach 1883: 578–9).[11] And so, in an attempt to cope with both the detail and the mysteries of our experience of the physical world, we are driven to generalise and to symbolise: a necessary compromise between the desire to understand all and the recognition of our limitations as human observers. It is a necessary 'sacrifice of exactness and fidelity'.[12] Mach sees that scientists should not waste their time unnecessarily on tedious procedures. They should economise as much as possible on their time and their efforts. And symbolic generalisations and abstractions help them to achieve this goal.

2 Mach's regard for mathematics is enormous. Without mathematics, the task of describing the physical world in a powerful way might indeed seem futile. The very forms of mathematics and geometry are amongst the greatest allies of science. And so scientists should be concerned to capitalise on the flexibility and power of these forms. Economy in the form of a description should be the goal of every scientist. Mach frequently cited Newtonian mechanics as an exemplary model for all scientific theories, given the elegant and economical form of its mathematical and geometrical descriptions.

3 Mach believes that scientists should be economical, not just with their labour and in the form of the descriptions they produce, but also with regard to the content of those descriptions. Science should aim to minimise its epistemological commitments: for example, if belief in three basic kinds of 'stuff' is sufficient to describe the world, then we should not add to those beliefs – if we suppose any additional entity to exist, then that supposition is likely to be no more than metaphysical speculation about the world 'beyond' the facts. Mach sees little to be gained by engaging in metaphysical games. He does recognise the importance of using terms like 'atom' for summarising key features of our experience. But use of such terms implies no epistemological commitment to unobservable entities as such:

they are merely shorthand terms standing for complex descriptions. So Mach urges scientists to wield Occam's razor to cut away superfluous entities from their essential commitments: the building-blocks of science should be kept to a minimum.

Although Mach's ideas about simplicity and economy blend well with his commitment to observation and our direct experience of the world, these ideas carry a wider, deeper methodological message for our approach to science. Scientific methods themselves should be constrained and enriched by the search for simplicity, which, Mach tells us, is the 'fundamental conception of the nature of science'.[13]

Mach's ideas on economy are no more than a sketch for a more detailed account. He fails to recognise that the various goals of economy in labour, form, and content might point us in quite different directions in trying to characterise the world.[14] A more serious problem is his failure to distinguish between subjective and objective ideas of simplicity.[15] But perhaps the most serious problem is raised by Michael Friedman, who argues that additions to our basic building-blocks, which Mach might regard as metaphysical extravagances, can sometimes play a unifying and simplifying role in our general accounts of the physical world.[16] Friedman sees the concepts of space and time as playing such a role, helping to bring together gravity and electromagnetism by referring all phenomena in these distinct domains to the same background framework. Although Mach certainly captures an essential feature of the scientific enterprise, his views on simplicity are just too broad and hazy to help us determine specific criteria for science and its methods. But his views do help us to appreciate his attack on absolute space and time as, at least in part, motivated by methodological goals of simplicity.

POSITIVISM IN ACTION

Mach is frequently regarded as the chief instigator of scientific positivism, a philosophy of science which regards the possibility of observational and/or experimental verification as the defining characteristic of all scientific statements. His empiricist polemic echoed the views of earlier philosophers such as Berkeley and Hume. And others during the nineteenth century voiced similar beliefs. Mach's influence in the scientific world was far-reaching.

Many scientists were initially excited by what they believed to be the possibility of freeing their domain once and for all from metaphysical speculation. They welcomed Mach's attack on Newton, as well as that on atomism. And they were impressed with his condemnation of those who are tempted to go too far beyond the phenomenal level of sense experience. Like Mach, they maintained that what cannot be observed by the senses cannot have any more than instrumental value in science. Many in the philosophical community were perhaps even more enthusiastic. They looked for ways in which to give rigorous expression to Mach's phenomenalistic philosophy of science. Empiricism, a vigorous anti-metaphysical approach, and a close acquaintance with language and logic were all combined to produce logical positivism – a philosophy according to which statements are meaningful only in so far as we can verify them (at least in principle) by means of our sense experience. From these ideas, we may distil three essential aspects of positivism programmes:

1. We should regard sense experience as the only admissible guarantor of our physical descriptions; hence statements involving an essential reference to theoretical or unobservable entities may have at best an instrumental status in our accounts of the world.
2. Our knowledge about the world may only be regarded as secure if it may be checked against observation and experiment.
3. We should not seek anything more than complete descriptive powers in our accounts of the physical world; 'fundamental' explanations, particularly those involving supposed causal connections or metaphysical entities, should have no place in science.

Apart from the lack of a positive theory for inertial interactions, two related objections to Mach's positivistic views may be made. First, his conviction that observation and experiment provide the only basis for science is suspect. For, as positivists have discovered, it is hard to make any clear distinction between observation and theoretical statements or between observable and unobservable entities. Secondly, his view that counterfactual suppositions should be constrained by observational evidence is problematic. We need a clear account of why our physical theories should be restricted in scope and Mach does not provide this. The first difficulty is more serious and how we respond to this will also determine our response to the second, since this too relies on the views taken both of the status of theoretical statements and of any distinction there may be between observation and theory.

Positivism has suffered many blows; but perhaps the fiercest is the claim that observation statements are theory-laden or theory-dependent. Given that Mach relies on the basic presumption that observation is epistemologically pure, this blow strikes him just as hard as the later positivists. If what we say about the world is always dependent upon our non-observational beliefs, then it may be hard to justify giving observational statements any preferred status.

Thomas Kuhn is one of many who argue that our observational claims depend upon our theories and beliefs. In *The Structure of Scientific Revolutions*, he discusses how scientific education eventually transforms a student's vision of the world in a profound and radical way and how revolutions in science produce similar '*Gestalt*' shifts in thinking. To illustrate his general claim about the way we see the world, Kuhn considers Wittgenstein's example of a drawing which might be seen either as a duck or a rabbit:

> What were ducks in the scientist's world before the revolution are rabbits afterwards. ... Transformations like these, though usually more gradual and almost always irreversible, are common concomitants of scientific training. Looking at a contour map, the student sees lines on paper, the cartographer a picture of a terrain. Looking at a bubble-chamber photograph, the student sees confused and broken lines, the physicist a record of familiar subnuclear events. Only after a number of such transformations of vision does the student become an inhabitant of the scientist's world, seeing what the scientist sees and responding as the scientist does. The world that the student then enters is ... determined jointly by the environment and the particular normal-scientific tradition that the student has been trained to pursue.
> (Kuhn 1970: 111–12)[17]

Kuhn's message is: what we see and certainly our reports about what we see are dependent upon us – upon our education, our social context, our culture, our general beliefs, our scientific theories. Because we can find many examples of different people in different places and times who would disagree (sometimes strongly) with even our most basic observational reports, there is no justification for the claim that what we say we see is correct. We cannot even say that two people from different contexts are looking at the same thing and that any difference of opinion about the object is simply about the precise description to be given. For we may find that the disagreement is over the very existence of the object.[18]

In the final two sections of this chapter, we shall consider the status of observational and theoretical entities and statements in greater detail. This will allow us to judge whether we have any firm empiricist grounds for repudiating Newtonian physics and its apparent commitment to space and time as absolute entities. In the next chapter, we shall consider alternatives to Newtonian theory. There we shall discuss the extent to which Mach's vision has been vindicated by twentieth-century theories.

CAN WE SEE SPACE?

The ideas of Kuhn and others have provoked three basic responses to the problem of the status of observation.

1 The spectrum view. Although we cannot make a firm distinction between observation and theory, we may nevertheless rank our scientific statements in a league table or view them as belonging to a spectrum: at one end the highly theoretical claims about such 'entities' as quarks and black holes and space, and at the other very basic observational claims about such immediate experiences as colour and length and motion relative to us.[19] We may then argue that we have every right to be more confident in our assertions at the observational end of the spectrum. For it is rational to have strong doubts about statements which are only indirectly and loosely connected with our day-to-day experience. But it is far less plausible to question this more direct experience – which, through its deep entrenchment in our belief system, becomes 'contingently incorrigible', i.e. although we may doubt our most basic statements, it is often difficult to advance convincing reasons for such doubts and therefore such doubts may not be altogether rational.[20] This approach gives us a reason for preferring a theory without references to theoretical entities such as space and time. But we have no justification for ruling out Newtonian theory simply because it does make essential references to space and time. So the best we can do is to

look for an alternative theory without such references.
2 All statements are theoretical. Paul Feyerabend says that even the most basic 'observational' claims are theoretical.[21] Our view of the world is not an essentially observational perspective which is then influenced by our theoretical beliefs. Rather, our view is theoretical in every way. Consequently, two people, from different theoretical backgrounds, looking at a scientific experiment would see not one experiment but two. Each observer would see meter readings and other such data from their own particular theoretical perspective. Even if they agreed before an experiment that the pointer of a meter moving in a specific direction would confirm one view and be anomalous in the other, the experiment would not provide an 'objective' test between the theories. If even the most basic data are bound up with the theoretical world-view of an observer, it would be a mistake to think in terms of a spectrum ranging from the relatively observational to the relatively theoretical. If we follow this more extreme line of argument, then the option open to us above (to search for a 'non-theoretical' alternative to Newtonian theory) is now closed. For there is nothing to be gained by adopting a theory which is less 'theoretical' in content.
3 We can see if *we* can see. Although we cannot clearly distinguish between observational and theoretical statements, we may maintain a firm distinction between observable and unobservable entities. Berkeley seems to be on safe ground when he says that Newtonian space and time are unobservable. But what makes Newton's argument so persuasive is the claim that space has observable effects. We might not see space directly, but we can 'see' it indirectly by the way it acts on material objects. Scientists often rely on such indirect approaches: from inferring the nature of the core of the Earth or of a distant planet from surface activity to using X-ray photography to detect 'unseen' tumours. In *The Scientific Image*, Bas van Fraassen argues that there is a natural limit to these indirect inferences. He says that, when I look at the vapour trail of a jet in the sky, even though it is out of sight the jet is observable because in the right circumstances I could observe the jet, e.g. if I were close enough. But, when I examine the trail of a micro-particle in a bubble-chamber, I cannot count the particle as observable because there is *no* circumstance in which I could observe the particle. The powers and capacities for observation possessed by human

beings determine what is and what is not observable. There is no way in which space and time may be observed by humans. 'Seeing' something via its effects, whether this is a trail in a bubble-chamber or an inertial force, does not amount to a case of observation unless we could in principle genuinely observe the thing via the senses which we happen to have.

Newtonian theory may involve an essential reference to space, but there is no necessity for us to believe all that the theory tells us: we may instead accept the theory on the basis of what it has to say about observables, of its 'empirical adequacy'. He says: 'Science aims to give us theories which are empirically adequate; and acceptance of a theory involves as belief only that it is empirically adequate' (van Fraassen 1980: 12).[22] Science, for van Fraassen, is primarily about observables. So acceptance of Newton's theory does not necessarily commit us to any belief in space and time *per se* since they are not observables in van Fraassen's sense. This seems a promising plausible approach for a Machian to take – it underwrites his/her observational bias; but it does *not* give the Machian a reason to look for an alternative to Newtonian theory. The Machian may simply *accept* Newtonian theory and be content that this acceptance does not commit him/her to a belief in the existence of space. Therefore, we might be concerned that any naïve acceptance of van Fraassen's basic approach could be a recipe for theoretical stagnation. We might also be concerned about van Fraassen's faith in our ability to provide an objective account of the powers and capacities of observing humans. He admits that this is an empirical problem. But he fails to recognise that, like all empirical problems, our resolution may depend upon our wider system of belief if not directly upon our theoretical prejudices. In other words, our account of observability may itself be theory-dependent.

EXPERIMENT AND INTERVENTION

Ian Hacking argues that the above talk of observational and theoretical entities and statements does not give us a satisfactory account of the nature of science. He says that three issues must be faced:

1 When we observe, we should not use just our eyes; we should not be mere spectators. The secrets of a microscope slide will only be

revealed to those who learn to intervene, to interfere with the specimen under scrutiny. Just to peer down the microscope passively will tell us little or nothing. We learn by experiment and experimentation demands active intervention – it is not a spectator sport. Therefore, a more complete picture of the scientific enterprise demands an account of the active relationship between observation and experiment.[23]

2 We must recognise the fact that observation is not so closely related to or dependent upon theory as some, including Feyerabend, have supposed. When I see an open door, my observation does not depend on theories about how light travels from the door to me. I need know nothing at all about the nature of light to feel confidence in my claim that the door is open. Of course, there may be a strange force-field which affects the light in such a way as to make a closed door appear to be open. But, since there are no such theories about such fields, my observations can hardly be said to depend on them. If theory-dependence is to make any sense at all, then a statement must depend on articulated theories. I simply do not have a theory about deep-sea creatures which influence the pictures on my television set by their extraordinary mental powers. So it would be foolish to suggest that my observation of a politician on the screen depends upon my denial of such a theory.

3 'Experimentation has many lives of its own'; for many experimental traditions survive quite happily without any theoretical basis. Hacking cites the history of superconductivity as providing an excellent example of this: 46 years passed before the experimental domain of superconductivity gained a theoretical base in quantum mechanics.[24] Experimental science can have an independent concrete existence with its own subject matter and assumptions, instruments, and data and the various operations performed on the data, together with a distinct social context for experimental scientists. Clearly theory, observation, and experiment do interact – but given the separate 'lives' they lead, these three domains may be best described as partially autonomous.[25]

Hacking regards the active intervention of the experimental scientist as the primary business of science. He recognises that theory is 'crucial to knowledge, to the growth of knowledge, and to its applications'.[26] But, if we want the firmest of grips on the physical world, we should not think in terms of theoretical representations. Too many problems face the realist who regards theories as candidates for representations of the world. A major problem, which we have already noted for this kind of realist, is that theories may be generally underdetermined by data; see Chapter 4. But the history of science also shows that theories provide no complete guarantee of stability. For references to unobservable entities like phlogiston sometimes disappear after a change of theoretical direction: and, if one reference can disappear, why should we have confidence that any of our theoretical terms refer to things in the world? The best that this brand of realist seems able to do is to ask for charity and to offer a principle of faith and hope: give me the benefit of the doubt – I am usually right![27] Hacking wants rather more security for the knowledge base of science. And he tries to persuade us to move away from representational realism to a realism about entities. The added bonus we get from this move is a high degree of stability for the scientific enterprise. He argues that it is the ability to manipulate entities like electrons in order to produce independent results in experimental contexts which underwrites our belief in the existence of the electron. We might be wrong in our descriptions of electrons, but we cannot be wrong about their existence. Thus Hacking asks us to be realists about entities, but sceptics about the theoretical descriptions of these entities. When we can manipulate an entity, we get a grip on something in the world which will not be disrupted through theory change.

So, like van Fraassen, Hacking views theoretical 'knowledge' as secondary in the scientific enterprise. With his 'positivistic' bias towards experimentation and his belief that low-level observation is free from theoretical prejudice, Hacking offers some comfort to the Machian. Space and time are not manipulable and therefore they are not 'observable' in Hacking's sense. And so we have no reason to believe that space and time actually exist. When we add to this Hacking's recognition of the importance of theory for the development of knowledge, we have at last a strategy which the Machian might adopt in a search for a consistent relationist theory of inertia and motion. But to do so requires a compromise. The cost for the Machian may be that they have to admit that entities do possess intrinsic causal dispositions and powers: for successful manipulation in a variety of contexts, using unobservable and observable entities alike, seems to commit us to the belief that these entities have definite causal characteristics, even if any particular description of these properties may itself be suspect. And Hacking, unlike the positivists of old, sees no virtue in restricting his attention to the low-level end of the observation-theory spectrum. Although Hacking admits to doubts about the status of black holes, unobservable or theoretical entities are not banished from science. For the history of science shows us that what is unobservable today may be manipulable tomorrow. Could we with some advanced technology manipulate space and time? I think we cannot rule that possibility out. And so, despite Hacking's positivistic commitment to observation and experiment, he provides a strategy which might dismay the Machian who is determined to repudiate space and time come what may. But the same strategy offers hope to the less dogmatic Machian, who sees the material world as the central foundation for our physical theories. If we grasp this world through intervention, and space and time turn out to be manipulable, then space and time would be aspects of this material world. I think Mach himself would have seen a glimmer of hope in Hacking's approach.

7

EINSTEIN AND ABSOLUTE SPACETIME

INTRODUCTION

The General Theory of Relativity (GTR) has frequently been regarded as the ultimate defeat for absolutism. For some time after the development of the theory, writers following the lead given by Hans Reichenbach continued to claim: that the ideas of Leibniz and Mach are vindicated by GTR; that Albert Einstein overcame the need for irreducible references to absolute space and time in our physical theories; and that inertial effects may be understood in terms of material interactions alone.[1] But a succession of contemporary writers have shown that such claims are not fully supported by GTR as normally understood.[2] In order to explore the issues involved, we need to focus clearly on the idea of a model of GTR. We need models to characterise a tremendous range of gravitational contexts, from planets orbiting the Sun to distant rotating black holes.

Relativists typically construct and analyse models of GTR – sometimes called solutions of the field equations – which consist in the following elements:

1. a four-dimensional geometrical background framework or 'manifold': spacetime – a continuum of spacetime points;
2. a matter 'field' which represents the distribution of matter and energy in a spacetime – moving beyond the Newtonian conception of matter as persisting, distinctive particles and invoking Einstein's idea that matter and energy should be interrelated in relativistic theories by the fundamental equation: $E = mc^2$;
3. a metric which has a 'flat' (Lorentz) character at least locally;

4 a set of field equations, with at least a strong similarity to the Einstein field equations of 1915–16, which relate the metric and the matter field, and which incorporates the idea of an affine connection that defines the idea of force-free 'straight-line' motion in the spacetime.³

The key to GTR is often said to be Einstein's use of the locally flat Lorentz metric, which allows us to decipher the complexities of gravitation and motion by reference to local inertial frames rather than to global frames. The local inertial frame provides us with a standard for inertial, force-free 'straight-line' motion. In the local limit, i.e. an infinitesimally small region, we may neglect most gravitational effects. Gravitation tends to be associated with the more 'global' curvature of larger scale regions in spacetime.⁴ The mathematical apparatus of GTR links local and global geometry and thereby guarantees that the geodesic of an object in curved spacetime corresponds to the locally defined 'straight-line' motion of the object.⁵ The idea of straight-line motion is defined in terms of the affine structure of spacetime, and in the local limit the affine 'connection' is zero, signifying the absence of forces. The affine connection is a geometrical quantity employed in GTR's equations of motion to describe the motion of particles in spacetime. See Chapter 3, Figure 11 (p. 54), for a brief reminder of the various levels of structure, including affine, involved in spacetime geometries.

Newton asserted that motion in our local environment may be referred to the global framework of space; Mach argued that the only admissible global framework is that of the fixed stars. Although Einstein refers motion to a local inertial framework in GTR, the intimate and essential relationship between local and global geometry allows us to ask two questions: specifically, whether the affine structure of spacetime and therefore our idea of inertial motion is determined unambiguously by the distribution of material in spacetime; and, more generally, whether the descriptions of global geometry may be reduced without remainder to material terms. Despite the hopes of relationists like Reichenbach, GTR may be used to construct models which do seem to involve an irreducible commitment to spacetime as an independent entity. There are a number of important models involving such a commitment: empty and almost empty models, including Minkowski spacetime and the Schwarzschild solution of GTR's field equations;

and models in which the material contents exhibit the same kind of overall rotation which Newton describes in his 'rotating bucket' thought experiment.

However, other models of GTR, including the standard Friedmann cosmological solutions of the field equations, do not seem to require any total absolutist commitment, at least when we focus on the problem of inertial motion. And we shall see that the champions of relationism in the twentieth century have tried to focus our attention on these, often claiming some pre-eminent empirical status for them. 'Non-absolutist' models are said to include: infinite worlds with an infinite amount of material; spatially closed worlds; and models with additional fields. Sometimes such models are generated entirely within GTR as characterised in standard textbooks, sometimes within variations of GTR. But, despite the ingenuity of those who try to realise the relationist dream, it is far from clear that all such models are free from an essential commitment to spacetime as a fundamental and irreducible entity in our explanation of the source of inertia. And, even if we can show that some models with a sound empirical pedigree do satisfy relationist demands, we still need to be shown why we should abandon all other 'absolutist' models, especially given that many of these have good empirical credentials as well.

MACH'S PRINCIPLE

The essential ideas behind Mach's Principle as used within the theoretical context of GTR derive from Ernst Mach's attacks on absolute space and time. We saw in the previous chapter that Mach believed strongly that science should deal only in observable phenomena and that our accounts of the physical world should be as economical as possible. Mach therefore saw no reason to embrace an entity which could not be seen and which, in his opinion, is an unnecessary and extravagant element in our general account of inertia and motion. Einstein was the first to refer explicitly to 'Mach's Principle' in a paper of 1918; he said that he chose the name 'because the principle implies a generalisation of Mach's requirement according to which inertia should be reduced to the interaction of bodies'.[6] In a paper written in 1917, Einstein found a static, spatially closed solution to his field equations, which he believed, for a short time, might be thoroughly Machian. But work by the Dutch physicist de Sitter demonstrated that Einstein's ideas were not fully consistent with Mach's Principle. We shall review Einstein's solution and his debt to Mach in Chapter 10.

Many other physicists, usually pursuing Machian programmes, have given their own versions of the principle. Wheeler gives a typical statement of the principle as follows: 'The geometry of spacetime and therefore the inertial properties of every infinitesimal test particle are determined by the distribution of energy and energy-flow throughout all space' (Wheeler 1964: 305).[7] Since we are dealing with the inertial properties of objects in motion, and these are determined by the affine structure of spacetime, it seems reasonable that we should account for this fact in any statement of the principle. So we might restate Mach's Principle as follows:

MP The affine structure of spacetime is uniquely determined by the distribution of matter/energy in the universe.[8]

This has two immediate implications: first, in an empty spacetime, there is no meaning to the idea of inertia – since, without material contents, there can be no affine structure, and inertial properties only make sense in terms of affine structure; secondly, the only meaningful idea of motion is motion relative to other material objects – hence the idea of an overall rotation of the material contents of the universe just does not make sense. Hence, if GTR gives a natural and straightforward meaning to either of these possibilities, we would have little reason to say that GTR incorporates MP. Much of the considerable debate about MP turns on just these two possible situations. We need to examine the models of GTR involving the above features before we may arrive at any considered conclusion about the status of the principle in GTR.

ABSOLUTELY, PROFESSOR EINSTEIN?

The source of much of the confusion about whether or not GTR vindicates relationism lies in the fact that there are several ways in which the word 'absolute' is employed within the context of spacetime theories. In some senses, GTR's spacetime is clearly not absolute: for example, in both GTR and the Special Theory of Relativity (STR) the time measured between two given events depends upon the spacetime path taken between those events – as we saw in the case of the 'paradox' of the twins in Chapter 2. And, breaking free from Newtonian beliefs, in GTR as in STR, we may no longer identify a unique time-slice which represents the entire universe at an instant to which all events near and far may be referred. But in other senses GTR does seem to be absolute: for example, the infinitesimal spacetime interval between two neighbouring points is invariant. Different observers might measure different spatial distances and different times between the two points, but they will always agree upon the spacetime distance, i.e. upon the infinitesimal spacetime interval. So, before we begin to explore the conflict between absolutism and relationism within the context of relativity, we need a much sharper account of the various ways in which a spacetime theory may be absolute.

We have seen that Newton's account of space and time involves the following claims:

1 space is a three-dimensional arena in which objects are located and events take place; no object and no event has any effect on space itself – hence, it is dynamically independent of all the dynamic events taking place within it;
2 time too is independent of all the events taking place in time – and it provides an independent and global temporal framework to which all events may be referred in the same way;
3 because of the independent natures of space and time, we may always specify the distances and times between events in an unambiguous way;
4 hence, we may give an unambiguous sense to the idea of simultaneity so that two distant events will be regarded as simultaneous regardless of the state of motion of the person who makes the judgement of simultaneity;
5 when an object experiences inertial forces, we may say that the object is indeed in motion relative to space itself, so that acceleration is an invariant quantity which cannot be transformed away by a change of reference frame.

Newton's position allows us to distinguish three distinct ways in which space and time may be said to be 'absolute':

1 space and time may exhibit an 'absolute' independence of objects and events;
2 space and time may be 'substances' of some sort with distinct 'absolute' invariant properties;
3 space and time may be required as irreducible and essential 'absolute' elements in our general account of motion.

Newton's ideas of space and time clearly involve all three senses of 'absolute' here: space and time are independent – not only of material objects, but also of each other; they are entities with well-defined invariant properties; and acceleration in Newtonian space may only be properly characterised with respect to space itself. So Newton's beliefs about space and time allow us to identify these three important senses of the term 'absolute':[9]

AE 'absolute' as independent entities or substances;
AP 'absolute' as substances possessing invariant properties;
AM 'absolute' as irreducible elements in our general account of motion and objects in space.

We should not think that every concept of Newtonian theory is 'absolute'. For example, it is certainly not true that *all* concepts involved in the Newtonian account of space, time, and motion are invariant. Simultaneity is a frame-independent, i.e. invariant, notion because, in Newtonian physics, two objects will be judged as simultaneous whatever the frame of reference in which we make a judgement of simultaneity. But the concept of velocity is frame-dependent. If an object is free from the inertial effects associated with acceleration, we are free to choose any inertial frame as the reference point for our calculation of velocity; and so the uniform velocity we ascribe to an object depends upon the frame to which we refer the motion – all we may say is that the object is moving uniformly relatively to this or that frame of reference. Hence, velocity in Newtonian physics is a 'relative' rather than 'absolute' (AP) concept. This follows from the key idea of Galilean relativity of motion which lies at the heart of Newtonian physics: the laws of mechanics apply in all inertially moving systems so that every mechanical experiment in every such 'force-free' system has precisely the same results. So we cannot find experimental evidence which allows us to point to some unique privileged inertial system and say: that system is really moving at some well-defined velocity or is really at rest with respect to space itself; and so we cannot single out a privileged inertial frame of reference which is at rest with respect to space itself.

If we are to characterise absolutism in a relativistic context successfully, we need to take account of the geometrical language of STR and GTR. In these theories, we move from talk of space and time to the idea of spacetime – and so our account of absolutism must take this into account. This move from space and time to spacetime need not disrupt our view of Newtonian theory, for it is a straightforward (if technical) matter to express Newtonian physics as a spacetime theory using the geometrical language of modern spacetime theories. If anything, such a move helps us to appreciate the similarities as well as differences between relativistic and classical perspectives since both are couched in terms of the same geometrical language.[10] One of the distinguishing marks of Newtonian theory is, of course, the fact that we may construct unique global time-slices through each point to which we may refer all events in the spacetime. So we may characterise the Newtonian 'separateness' of time from space by pointing to the appropriate features of spacetime geometry which allow us to construct such time-slices. Consequently, there is much to be gained by setting discussions of absolutism in the context of spacetime theories in general, since the discussion will then apply equally well to relativistic and non-relativistic theories.

However, we should be cautious about using spacetime rather than space and time when we consider historical case studies. For example, if we wish to provide an accurate interpretation of Newton, we should not translate his idea of one-dimensional time 'flow' into a four-dimensional 'static' spacetime context. Newtonian time as such is uniform, fixed, and certainly flowing in some sense; but it is not part of a spacetime framework. If we could bring Newton back to life, he might well be persuaded that the Scholium and the *Principia* are inadequate when faced with the elegant and powerful mathematical formalism used by twentieth-century relativists, but the historical Newton knew nothing of spacetime as such. We can rewrite Newtonian theory in a modern 'spacetime' way; but we cannot rewrite history.[11]

Clearly, the most important sense of 'absolute' for someone who wishes to assess the role of spacetime in our account of motion is the third (AM). And this sense must also provide the focus for any relationist attack on absolutism about spacetime. For, if *all* talk of spacetime could be reduced to material terms, then we might say that all apparent properties of spacetime are in fact reducible to properties of matter and that spacetime cannot be genuinely independent in any way since it is no more than its contents. Therefore, the apparent independence of a spacetime and whether or not a property is invariant may ultimately be traced back to material factors. Consequently, the relationist is not immediately threatened by the facts that some properties of spacetime are invariant and that spacetime is at first sight an independent entity. We might therefore characterise the strongest relationist position as one which opposes the key, third sense of 'absolute'. A relationist, perhaps accepting Machian demands for an account of the world in purely observational and material terms, might therefore say:

RM all spatio-temporal elements involved in our account of motion are wholly reducible to material terms.

A natural consequence of RM is the more restricted idea that those spatio-temporal elements involved in descriptions of the *inertial* properties of matter are uniquely determined by the material distribution, i.e. RM implies that Mach's Principle (MP) holds. Hence, if we can show that MP does not hold in GTR, then it is clear that RM cannot be true of GTR. However, we should note that even if we can demonstrate that GTR in some sense incorporates MP, we may still fall short of showing RM to be true, for there may be more to spacetime than those elements involved in descriptions of inertia, e.g. topological elements.[12] Hence, if a Machian wishes to give a wholly material account of *all* the properties of spacetime then s/he would need to show that RM holds.

We should note that STR's Minkowski spacetime is typically empty of all material other than 'massless' test particles. But a particle moving inertially through this spacetime has its 'straight-line' motion determined by the spacetime: it will travel along a geodesic where the spacetime intervals along the geodesic are determined by the affine geometry of the spacetime. The spacetime interval is an invariant property of STR's spacetime, since where the particle moves next in spacetime is a fact which may be agreed from all points of view, from all frames of reference in the spacetime. There is no other matter in the spacetime; so an irreducible reference to the geometrical properties of the spacetime is required in a complete description of the particle's motion. Therefore, in STR, we can talk about the motion of a test particle without reference to any other material in spacetime with perfect sense. Hence, RM cannot be true of STR. We treat STR's spacetime as a substance with specific geometrical properties. Although its properties are clearly different from those of Newtonian space and time, STR still incorporates geometrical structures which underwrite such absolute (AP) concepts as acceleration and the spacetime interval. And since STR's spacetime structures are unaffected by any material in the spacetime, we may also maintain that AE is true of STR.

GTR certainly offers more hope to the relationist. For Einstein's field equations of 1915–16 reject the idea of a spacetime which is independent of its material contents. Matter, understood in terms of a mass–energy field, and the geometry of spacetime are related in a direct way: the distribution of matter influences the overall affine and metrical structures of spacetime; and, conversely, the geometry of spacetime determines the paths taken by inertially moving particles. When we speak of matter producing curvature in GTR spacetimes, we usually mean curvature of the metric; and, of course, measurements in curved spacetime typically differ from those in flat spacetime. But a curved metric also indicates that the affine structure has been affected by the presence of matter. The effect on the affine structure is particularly important when we consider the problem of motion, given that the affine structure of spacetime allows us to characterise the 'straightest possible paths' which force-free particles follow in spacetime. Hence, in GTR, spacetime is not a completely independent entity, i.e. it is not AE, at least when we think only of affine and metrical structures. However, we must remember that other 'deeper' structures are involved in a full description of any given spacetime: for example, we need to take account of topological properties. Therefore, the fact that GTR involves an explicit relationship between affine and metrical structures and matter has no immediate implications concerning the topological independence of spacetime in GTR.

The relationship between matter and geometry gives some hope to the relationist who wishes to reduce spacetime to material terms. But this dynamic relationship provides merely a possible mechanism for the reduction. We may nevertheless require spacetime concepts in our general account of motion. Unless we can demonstrate that a sound theory of spacetime and motion such as GTR may be spelt out *fully* in material terms, relationism seems destined to be no more than a pious empirical hope.

EMPTY, ALMOST EMPTY, AND ROTATING WORLDS

It might seem empirically unsound to consider empty models of spacetime theories as having any sensible implications at all for our view of the 'real' world. We might just dismiss them as 'unrealistic': if our world is anything, it does not seem to be empty!

However, the relationist cause is not particularly helped by taking up some strong Machian line and saying that we should take seriously only those models which correspond closely to the actually observed universe. For the actual observations seem to be uncertain enough to give cosmologists a tremendous amount of leeway. We can easily imagine a universe, resembling our own in many ways, which effectively amounts to an empty world. If there is a finite amount of matter in the universe congregated at the 'centre' of some vast infinite arena, then this world, considered from a truly global perspective, approximates to a system in which a single mass is located in an otherwise empty infinite spacetime.

Although present cosmological evidence suggests that the universe does not have this particular structure, the uncertainty of this 'locally' gathered evidence should be sufficient to prevent any hasty dismissal of the idea that this effectively empty world is a possible scenario for the actual universe. And, if we consider the single mass in an otherwise empty universe as a particle without gravitating properties and we focus on non-gravitational aspects of motion, then even the empty Minkowski spacetime of STR may be considered as a plausible empirical model for the universe. In fact, some relativists view the 'flat' spacetime of STR as a 'natural' geometrical background for material.[13] The introduction of a finite amount of gravitating material into this spacetime produces significant curvature locally (and therefore takes us into GTR's theoretical context), but far away from the material the curvature effects are negligible and, at infinity, are zero. Globally such a spacetime is effectively 'flat'.

We have already noted that any account of motion in Minkowski spacetime involves an irreducible reference to the geometrical properties of spacetime. In STR, the affine structure of spacetime determines how any free particle may move; but we may not explain the affine structure in material terms. But Minkowski spacetime is just one of the many models of GTR, a 'special' case in a 'general' theoretical context. Hence, at least one model of GTR *seems* to be absolute in every sense that STR is absolute.[14] But, strictly speaking, Minkowski spacetime, as a model of GTR, should not be considered as an entity which is independent of its material contents. GTR's field equations are used to construct the model, and so the geometrical terms in the equation describing the affine and metrical properties of its spacetime are determined in part by the terms describing the distribution of matter. The fundamental relationship between matter and geometry expressed by the field equations cannot be ignored just because the terms describing the matter distribution indicate an empty spacetime. Hence, all models of GTR are relational in the sense that spacetime is not an absolute independent entity, at least in terms of its affine and metrical properties, i.e. they do not imply AE at these levels of structure. They may, of course, still be absolute entities or substances at a topological level.

The Schwarzschild solution of the field equations supplies us with a standard model of GTR which may be used to approximate a variety of physical situations, including the solar system. A single spherical gravitating mass is located in a spacetime with spherically symmetric spatial properties. Test particles in this spacetime may be used to represent the behaviour of objects moving close by the central mass, which may itself represent the Sun. The central mass produces marked local curvature in the metric of spacetime, and so any freely moving test particle will follow geodesics in this curved spacetime. And so we may use the model to predict the behaviour of objects in the vicinity of the Sun. However, at considerable distances from the central mass, the gravitational influence on test particles is negligible. How they move no longer depends upon the gravitating properties of the central mass, for, at great distances from this mass, spacetime behaves as if it were empty. The spacetime paths of such test particles are determined by the overall spacetime geometry. This is the case in all such 'almost empty' spacetimes – where a finite amount of matter is located in an otherwise empty infinite arena.

However, the overall geometry of a spacetime model is not, in general, fixed even though we specify the material distribution. In order to determine the affine structure, we must also specify the boundary conditions which are to obtain in the model. For example, we might say that the spacetime in which material is located is spatially unbounded – infinite in the sense that objects may be located at infinite distances from each other; or we might say that spacetime is spatially bounded so that space may be thought of as a three-dimensional analogue of the surface of a sphere. If we allow infinitely extended spacetimes with a finite amount of material in them, as in the Schwarzschild solution, then in general the paths of objects are effectively determined, not by the presence of matter in some local environment, but by the infinitely extended structure of the almost empty spacetime. We impose an overall geometrical structure on the spacetime when we specify the boundary conditions, and this structure is then essentially and irreducibly employed with definite metrical and affine properties in our account of motion in the spacetime. Therefore, almost empty and spatially unbounded spacetimes are, like empty spacetimes, absolute in the sense most threatening to the relationist programme: they are irreducible elements in our account of motion (AM).

The second major problem for relationists concerns the possibility of global rotations in spacetime. If the entire matter content of the universe is rotating, then the rotation could not be with respect to some non-rotating material body, for everything is rotating. The rotation could only be with respect to some element of spacetime itself, and so we could only make sense of the motion of material with an irreducible reference to some element of spacetime itself. Do such global rotations make sense in GTR? There are several models of GTR which seem to allow the possibility of an overall rotation, including:

1 Gödel's solution, in which matter is represented as a perfect fluid throughout the spacetime and in which space is rotationally symmetric about any point in the spacetime;
2 Kerr's solution, describing the properties of an isolated, rotating mass; and
3 the Oszvàth and Schücking solution, which characterises a globally rotating, homogeneous, and spatially closed dust-filled spacetime.[15]

The relationist might find it easy to be suspicious of the first solution above; but it is more difficult to challenge the status of the other two. Gödel's solution involves the possibility of closed loops in time, which, as we shall see in Chapter 8, may well commit us to physically implausible situations, if not outright logical contradictions. So we might be justified in adopting a severely sceptical attitude towards Gödel's solution. But, if we are to describe the properties of an isolated rotating object, then we clearly need to use some model such as Kerr's solution. And, apart from the rotation which it allows, the Osvàth and Schücking solution should not be immediately objectionable to the relationist since its spatially closed boundary conditions avoid the kind of problem involved in almost empty, spatially unbounded solutions where spacetime conditions at 'infinity' determine the way objects move in spacetime.

RELATIONISM AND RELATIVITY: AN EMPIRICAL VIEW?

Despite the number and variety of models of GTR which seem to involve an irreducible reference to spacetime itself, there are several models which seem to be generally in line with the relationist programme. The most celebrated of these are the Friedmann cosmological models, initially developed by the Russian mathematician Alexander Friedmann in the early 1920s.[16] Friedmann uses the Einstein field equations to construct a set of equations which characterise a variety of homogeneous and isotropic dust-filled universes. These models have a high empirical status, given that they allow the possibilities of global expansion and contraction of the material contents of the universe in both spatially open and closed contexts. They do not exhibit any overall rotation and their affine and metric structures are uniquely determined by the distribution of material in them. Hence, they seem to satisfy MP.[17] But they do not satisfy RM because the topological properties of the spacetimes are not uniquely determined by the distribution of material. As we noted in Chapter 3, the conditions of homogeneity and isotropy for, say, an expanding matter-filled universe do not force us to adopt a spatially unbounded topology: we may decide to adopt the topology of a torus instead, as is done in the case of the 'small universe'.

The empirical pedigree and the Machian character of the Friedmann models have encouraged some relativists to use them as a focus for their relationism. Following work done by Dennis Sciama and others, Derek Raine shows that we may construct a Machian variant of GTR which identifies a limited number of solutions of GTR in which two conditions are met: first, if we specify curvature, then a unique metric should be determined; and, secondly, the Weyl component of spacetime curvature should be determined uniquely by the matter distribution.[18] Together the conditions ensure that the matter distribution determines curvature, which in turn determines the metrical and affine structure. The equations which he employs differ from GTR inasmuch as they allow us to sum the effects of distant matter on any given material object: only linear theories allow this integral approach and GTR is non-linear. But Raine's modified equations do have the 'usual' field equations as a limiting value. And, given our remarks above, it is perhaps no surprise that the main group of

143

solutions, consistent with Raine's Machian constraints, is formed by the Friedmann models. Empty, almost empty, and rotating solutions are all ruled out by the conditions. Although Raine's approach is certainly inventive, his strategy has much in common with the approach of other relationists who stick with the usual field equations. Mach's Principle is used as a criterion for the selection of 'permissible' models of GTR. The advantage of Raine's integral formulation of GTR is that it gives us clear, if complex, mathematical guidelines on how the criterion is to be applied. But we still might decide that the overall strategy here is somewhat *ad hoc*: we will only admit those solutions in which MP holds; and, if GTR itself does not provide sufficient reason for us to throw out undesirable models, then we shall adopt a more obliging variant of the theory.[19]

The Machian relationist tries to force our attention towards a limited number of global cosmological solutions, perhaps allowing other models a merely instrumental status in our general account of inertia and motion. We may use models in which MP does not hold, but we should not expect such models to apply to the real world. The Machian sees no problem in this, for the approach is justified by a strong belief in the primacy of observation. We do not see an empty universe, so Minkowski spacetime carries no implications about the structure of the real world. We do not see an isolated mass in an otherwise empty universe, so the Schwarszchild solution has limited value also. We do not observe any global rotation, so the Oszvàth and Schücking model does not apply to the actual universe.[20] We do see a homogeneous and isotropic matter-filled universe, so any model which fits these facts is potentially acceptable. Hence, the relationist narrows the available options consistent with GTR's field equations down to those solutions which correspond with the observed conditions in the physical world. But the range of empty, almost empty, and rotating solutions should certainly not be regarded as 'second-class' models: without them we cannot make sense of such local irregularities as the solar system!

The incorporation of MP into the gravitational context of GTR seems to demand a rather cavalier attitude towards the generality of the field equations. We are asked to reject a rich variety of solutions on the grounds that they do not correspond with actual global observations. The fact that certain aspects of a model might apply to a local situation does not give us sufficient reason for accepting the model as empirically valid. If matter everywhere determines what is happening in any given locality, then the only solutions which carry any significant weight will be those which characterise, at least approximately, the actual global matter distribution. Consequently, the Machian relationist seems to require that the field equations should have a limited range of application, defined by the available empirical evidence.[21] But many scientists regard the major strength of physical laws as their generality, their applicability to all possible physical situations. This kind of generality underwrites Newton's argument for absolute space in which he feels free to apply his laws to an almost empty universe. And it is this unrestricted generality which Mach resists when he asks us to focus on the actual universe and not some imaginary artefact. Although Mach certainly allows that laws should apply in a reasonably broad range of circumstances, he would no doubt agree with Hermann Bondi's assessment of general relativity and the range of applicability of its laws:

> I take an extremely empirical view of general relativity. It is a theory like all other physical theories, founded on experiment and observation. We expect a decent theory to go a *little further* than the experiments and observations on which it is based, to account for something a little more general than the particular circumstances in which the theory became established. But we never expect our theories to hold in circumstances utterly and completely different.
> (Bondi 1967: 75–6)[22]

The major problem for empiricists like Bondi is how to draw the line between the empirically acceptable and the unacceptable. Relativists are indeed only too willing to rule out some logically possible situations as highly implausible. For example, Hawking and Ellis specifically restrict the admissible solutions of Einstein's field equations to those 'exact solutions' which satisfy two empirically based conditions: first, all causal signals in admissible solutions must travel along non-spacelike paths; and, secondly, local kinetic energy should never be negative.[23] We have every reason to suppose that these conditions hold for all known forms of matter.[24] But these conditions still give relativists tremendous flexibility in their construction of models of GTR.

The relationist might now say that we have every right to impose further, tighter restrictions in line with observational evidence. But the evidence itself suggests that this relationist ploy cannot do the job they ask of it. When Mach asked us to recognise that we cannot neglect the background material reference frame of the fixed stars in our account of motion, everyone thought that the stars really were cosmic fixtures, that the universe is globally static. But the evidence of astronomers, from Edwin Hubble's discovery of red shift in 1928 onwards, strongly suggests that nothing is fixed in a dynamic universe which is evolving from something like a 'big bang'. There is still considerable debate about the conditions in the very early universe, and there is some debate about the likely 'final' state of the universe. The empirical evidence is simply insufficient to rule out all of the 'absolutist' models of GTR as physically implausible. For example, we have no reason to rule out a global rotation in the initial turbulent conditions of the 'big bang'; as the universe evolves, the rotation slows. This is at least as physically plausible as many of the weird and wonderful tales which are told about the earliest of times.[25] Hence, the observational evidence for an evolving universe requires that we take a flexible rather than restrictive attitude towards the physically admissible models of GTR. We should not dismiss the possibility that, at some time in the past, conditions were, as Bondi says, 'utterly and completely different' – or at least different enough to make the relationist Friedmann models redundant. Of course, it might turn out that the early universe turns out to be a Machian kingdom. But, without hard evidence, we have little reason to use MP as a criterion for selecting the only admissible solutions of GTR.

THE HOLE ARGUMENT AND SPACETIME POINTS

So far the initiative seems to be with absolutists. However, one of the devices they rely on may be turned against them. Absolutists remind us that GTR allows the construction of solutions in which the distribution of matter and energy does not determine the metric geometry unambiguously. They then argue that, because of this, MP and therefore RM fail. For they say that there is clearly more to spacetime than may be expressed in material terms alone. But the absolutists must face the hole argument, originally used by Einstein in the genesis of GTR, and now refined by a number of writers worried about the status of absolutism. Here I shall focus on the argument as developed (technically) by John Earman and John Norton and (rather more informally) by Paul Teller.[26]

GTR is typically presented by many relativists as a classical 'deterministic' theory, in which (in principle) the equations allow us to predict and retrodict with certainty, so long as we have sufficient information to feed into the equations.[27] Imagine that we know everything we need to know to construct a 'global' time-slice of spacetime – all space at an instant, remembering that such surfaces of simultaneity are not generally unique in relativity: i.e. we know all the facts about the physical conditions on the slice. Is this information, together with the field equations of GTR, sufficient to allow us to construct a complete history of the spacetime, i.e. to determine the past and the future of the slice?[28] However, as we shall see, it is not quite so easy to fix the history of the slice in GTR, and this problem takes us to the hole argument.

One of the main features of GTR is that its equations are generally covariant. GTR allows me adopt any coordinate system. Such a system provides me with a way of deciding distances and directions between objects. But, if I choose another frame, then the equations of GTR should remain essentially the same when I apply them to this new frame and system. If the equations remain essentially the same as we apply them to different frames of reference, then they are generally covariant. The equations of GTR are written in *tensor* form, because tensors have the required property of general covariance.[29] Although general covariance is a purely formal requirement, we should note that tensors describe physical states of affairs *geometrically*.[30] So the use of tensors commits us to a fundamentally geometrical perspective on the physical world. The advantage of general covariance in GTR is that we can adopt any coordinate system at all to describe the physical situation in the spacetime framework or manifold. However, because distances and directions will depend upon the particular coordinate system chosen, the use of generally covariant equations in GTR to capture any coordinate system whatsoever implies that the specification of distance and direction between points is not fixed but variable.

Let us imagine that we have just made such a choice of coordinate system. May we change coordinate system *without* affecting the physics of the spacetime? General covariance guarantees that we may do just this. But may we make changes to the distribution of material in a specific region (or 'hole') in spacetime so that the result is a new spacetime but one with the same physical *description* as the original? General covariance also allows this! We may construct alternative spacetimes with no immediate physical differences simply on the basis of the formal properties of general covariance.

Clearly, the possibility of producing two physically equivalent worlds assumes that we may pick out individual spacetime points. Teller calls the new world a 'Leibniz alternative': for the identity of indiscernibles seems to apply to the original and its alternative.[31] Each of these worlds has the same material content and the same relations amongst objects. They seem to be indistinguishable at a physical level. But they differ in terms of their attachment to the points of the spacetime manifold.

At first sight, this might appear to be yet another argument for some form of absolutism. Once again, there seems to be more to spacetime than may be specified in material terms. Here the spacetime structures we are encouraged to embrace are at the topological level. But absolutists must hesitate before accepting this argument. For the Einstein 'hole' story as told by Earman and others seems to involve the downfall of determinism: we may no longer be sure about the future even when we know all there is to know about the history of spacetime up to now. For different futures may be generated from a single set of information about the world despite the fact that the equations used are essentially deterministic.

Imagine that a spacetime slice contains a small region or 'hole'. GTR is rich enough in structure to enable us to construct different models, with exactly the same distribution of material on the slice, but with different topological features. We may do this by altering the way the physical fields attach to the points in spacetime to allow the construction of distinct models. This procedure allows us to build, on the basis of information given, (for example) two distinct models with exactly the same pasts and presents (as defined by the time-slice) but with different futures.[32]

Manipulations of points in the spacetime manifold certainly make sense in abstract mathematical terms. But must we commit ourselves to the belief that the spacetime manifold is a 'substratum' with an independent 'absolute' reality, i.e. that spacetime points have a 'substantival' reality over and above the metric of spacetime and its material contents? If talk of spacetime points is not to be understood literally, then we need not draw any conclusions about physical reality on the basis of any manipulation of those points. This is the path that relationists might be inclined to take. They might simply say that the only significant features of spacetime are at the physical level: so we should only talk in terms of material objects and their relationships to each other. They might then argue that the 'indistinguishable' worlds are not really two worlds at all: the identity of indiscernibles operating at the physical level guarantees that we are dealing with a single domain.

What does the hole argument leave for absolutists? As yet the issues are complex and unresolved. Some writers believe that we may legislate against changes at the topological level by appeals to the essential properties of spacetime points; others argue that, when spacetime points and relations are properly understood, we would find no real distinctions between apparently different worlds constructed by appeals to the way fields may attach to specific regions in spacetime.[33] Finding these substantivalist manoeuvres unpromising, Earman suggests that the way out of the dilemma may be to accept that absolute accelerations take place but to deny that the inertial structure upon which such accelerations depend is not embedded in a substratum of spacetime points as such. In other words, he believes that Sklar's idea of absolute acceleration without absolute spacetime, according to which acceleration is regarded as a primitive property or brute fact about physical bodies, may turn out to be the best option for the absolutist. Sklar's approach has the advantage of being at least a partial compromise between absolutism and relationism. Absolutists may hold onto their claim that acceleration effects may only be captured through an absolute inertial structure; and relationists may retain their belief that talk of spacetime points has no significance for the actual world.

Earman traces the debate between relationists and absolutists from a problem about inertial forces to worries about determinism. He says that he wants determinism to be given a fighting chance and so he challenges substantivalism. However, the force of his argument depends upon his decision to set the dispute between absolutists and relationists at the ontological level. Clearly, if what is at stake is the reality of spacetime points, then the hole argument should make the absolutist hesitate before embracing substantivalism. However, those absolutists who take a more instrumentalist attitude towards spacetime theories need not be embarrassed by Earman's attack. They might agree that a theory such as GTR involves spacetime structures – from the affine connection to the differentiable manifold of points – as irreducible features. But they might see no reason to add any claim about the reality of these structures. Earman encourages us to accept that absolutism with and absolutism without substantivalism are the only viable positions for those with anti-relationist sentiments. So long as we concentrate on the problem of absolute spacetime as an ontological problem, there is some justification for Earman's position. However, we may also view the absolutist-relationist controversy as a dispute about the extent of our epistemological commitments.

We may therefore identify a further 'instrumentalist' position in which the absolutist (a) avoids discussion at the ontological level and (b) argues that talk of spacetime *within a given theory* may not be reduced to talk of objects and events. i.e. that spacetime is an epistemologically essential structure *in that theory*. Indeed, many physicists involved in GTR, e.g. Roger Penrose, work confidently with spacetime structures as essential elements in their accounts of motion but nevertheless express doubts about the ontological status of those structures.

The absolutist who endeavours to keep discussion at the epistemological level need not share Earman's concerns about embracing substantivalism and thereby repudiating determinism. For any such repudiation would be relative only to a given theoretical context, whether this be GTR or some other spacetime theory. So, if we were to treat the spacetime of GTR 'substantively' – as an essential artefact but not a real substratum, we would not be committed once and for all to indeterminism. Whether or not to embrace substantivalism and therefore indeterminism in GTR would be weighed against other considerations. And, as we shall find in Chapter 10, there does seem to be a good deal of support for an indeterministic approach within the general context of GTR, given the fact that modern spacetime theories must deal with a range of essentially indeterministic phenomena such as black holes, naked singularities, and the early cosmos.

8

TIME TRAVEL

INTRODUCTION

When H.G. Wells wrote *The Time Machine*, he related the astonishment of his time traveller as the fateful lever is depressed, the machine is set in motion, and the future unfolds its secrets:

> I seemed to reel; I felt a nightmare sensation of falling; and, looking round, I saw the laboratory exactly as before. Had anything happened? For a moment I suspected that my intellect had tricked me. Then I noted the clock. A moment before, as it seemed, it had stood at a minute or so past ten; now it was nearly half-past three! . . . I pressed the lever over to its extreme position. The night came like the turning out of a lamp, and in another moment came tomorrow.
>
> (Wells 1958: 20)[1]

Many of us have been introduced to the fantasy world of time travel by Wells' masterful narrative. But do we have good physical reasons for accepting such exotic stories, or should we treat time travel as a mere fiction? Can we explain any alleged example of time travel in terms of our 'normal' views of events in space and time? Would we be forced to drop the common belief that causes must precede their effects? Does the idea of time travel involve logical or physical contradictions? Indeed, many writers regard the possibility of logical contradictions as the most convincing reason why we should regard time travel as an absurd fiction: if we can gain access to our past and change it, then we might even be able to kill all our great grandparents in their youth! But, if we did commit such a crime, we would not have been born so how could we exist to go back in time and commit it? However, the microscopic laws of physics are certainly indifferent to time direction: given full information about a region of space at a given time, we may use the microscopic laws of physics to show us both the future and the past of that region in an even-handed way. So we might think, from a microphysical point of view, that our reluctance to consider the possibility of travel backwards in time is a peculiarly human affectation arising from our limited macroscopic perspective.

Despite the fact that talk of space and time does not appear to be reducible to talk of material objects and the relations between them, our ideas of time, change, and causality are intimately connected. So, although we might acknowledge the independence of space and time, it still seems hard to discuss ideas of time and the direction of time without reference to both change and causation in the material world. Nevertheless, by thinking in terms of the basic concepts of 'earlier' and 'later', we do seem to be able to discuss 'the direction of time' in a way which, if not independent of objects altogether, gives us a fundamental grasp of the properties of time. D.H. Mellor presents such a position in his book *Real Time* when he says:

> The direction of time is the difference between *earlier* and *later*; a difference not easy to pin down. *Earlier* and *later* are not differently related to each other, for each is just the other's converse. . . . I have remarked how we can often see that one event is earlier, not later than another. To revive the old example, seeing a clock hand move clockwise is *inter alia* to see it pass '1' just before, not just after, it passes '2'. Being earlier in this case is a perceptible relation between events, as spatial distance often is between things. It no more needs defining than colour does: we just see it.
> (Mellor 1981: 140–1)[2]

As we observe events and things in time we observe variations, and we talk of such variations using the language of change and causality. Changes are changes in things, and change involves causation. Although events do not change *per se*, we still may speak of events as being causes and effects of other events.[3]

The notion that a cause cannot occur later than its effects seems to be deeply embedded in the way we think about the world. Does the direction of time determine that of causation? If it does, then the idea of backwards causation as such would be untenable. For

TIME TRAVEL

we could then stipulate by definition that causes could never occur later than their effects: causes would always occur earlier than their effects. The direction of causation would be unambiguously future-directed. Of course, we might nevertheless be able to reach points which lie to our past, but *only* by following a future-directed path: this might occur in those universes which permit closed timelike or null curves, e.g. in a cylinder world.[4] So, although we might be able to rule out backwards causation *per se*, we might still allow the possibility of time travel to the past via the future along closed 'loops'. However, if causal direction is independent of time direction, then we might have to grant the possibility of backwards causation as well. In this chapter, we shall assess the cases for and against both backwards causation and travel along closed timelike paths. But we shall also meet other, perhaps more exotic, variations upon the theme of time travel arising principally from ideas in relativity.

The conceptual apparatus of the General Theory of Relativity (GTR) allows us to articulate two main senses in which 'time travel' might be said to take place, in addition to the ideas of tachyons and spacelike travel outside the light cone discussed in Chapter 3:

1 backwards time travel and causation: by moving backwards in time between two points, allowing the points to be connected by 'causal' signals travelling at speeds less than or equal to that of light; and
2 closed timelike or null curves: by moving forwards in time but eventually reaching a point to the past of the starting-point, so that a closed 'loop' in time may be formed; since tachyons might also form such loops, at least in principle, we might usefully associate their effects with this kind of time travel.

However, we shall find that many scientists and philosophers strongly object to the general idea of time travel. There are three main kinds of objection:

1 the idea involves physical assumptions which are highly implausible and may therefore be ruled out on empirical grounds;
2 time travel may involve the postulation of logical contradictions and may therefore be ruled out on rational grounds; and
3 any story of travel into the past may be retold in terms of 'forwards' travel.

So, in this chapter, we shall review the two main candidates for adoption as genuine cases of time travel; and we shall see how they fare against the various objections which have been raised before passing a final judgement on the case for time travel. Before looking at the various possibilities in detail, we shall first discuss some of the features of spacetime structure which have a bearing on the problem of time travel.

SPACETIME STRUCTURE

Spacetime is generally well behaved. We do not see strange sequences of events which might only be explained in terms of backwards causation. But we also do not seem to encounter glaring examples of time dilation in our low-speed 'Newtonian' lives. Why should we be inclined to rule out causal anomalies but accept such a strange phenomenon as time dilation? Perhaps because we seem to have a deeply entrenched belief that time and causation march together in one direction. But is this belief justified? In order to discuss causal behaviour in any spacetime, we need to fill out the structure of the spacetime – and each detail we add to the structure acts as a constraint on the events and phenomena which might occur in spacetime. We can assess the case for each element of structure and in so doing provide an analysis of the relationship between spacetime and causation. Then we shall be in a position to clarify the ideas of backwards causation and closed causal loops and to make at least a preliminary judgement on their likelihood.

The following elements are typically regarded as fundamental parts of the causal structure of spacetimes in STR and GTR:

1 *Time-orientability*: the least we should be able to do in a spacetime is define the idea of a direction of time in an unambiguous and continuous way at any given point; a spacetime is said to be time-orientable when we can divide 'arrows' of time assigned to the point into two groups – those pointing (i.e. oriented) forwards and those pointing backwards; hence, the time direction at every point of spacetime may be fixed with reference to these two groups.
2 *Time direction*: we may then choose a particular direction at any given point, perhaps relying on thermodynamic or other physical considerations; for example, we might believe that the general tendency of systems and structures to break down with a

consequent increase in entropy (often called an increase in 'disorder') sets the direction of time's arrow, and so all decisions on time direction may appeal to such considerations; or, following Roger Penrose, we might appeal to the initial gravitational conditions at the big bang as the explanation for the direction of time.[5]

3 *Precedence and causal precedence*: we now allow the possibility of curves or paths between points in spacetime and we define a temporal order on the curve; if we wish to rule out such exotic possibilities as tachyons from the spacetime, then we must stipulate that only non-spacelike curves may have points on them which are temporally ordered; a non-spacelike curve between two points may then begin at one point 'p' and end at another 'q' such that p precedes q. When we allow causal signals to follow such non-spacelike paths in spacetime, we may then speak of an event at p causally preceding an event at q. The causal future of any given point p is then the set of all events causally following p, i.e. which may be connected with p by non-spacelike curves beginning at p and directed to the future of p. Similarly, the causal past may be defined in terms of the set of all points causally preceding p.

4 *Light cones*: given the (empirically plausible) assumption that the speed of light in a vacuum is the same when observed from any inertial frame of reference, we may introduce the idea of two light cones spreading out to the future and the past from an observer moving through any given point in spacetime; these cones may be defined by the paths that light rays take through that point: one cone contains its causal future, and the other contains its causal past.

We should note that element 3 rules out the possibility of backwards causation, i.e. of a causal signal travelling backwards along a non-spacelike curve from future to past. This prohibition will be the subject of the next section: 'Back to the past'. But we do not thereby rule out the possibility of closed causal loops. This is because so far we have only defined the direction of time in local terms. Locally the past is always past; but globally some curve may eventually find its way to the past *via the causal future* along a non-spacelike causal curve. However, there is no question of the normal temporal (and therefore causal) order of precedence being reversed: each new event to be encountered lies to the future.

Nevertheless, such curves are clearly anomalous, in the sense that by following the curve around the closed loop we may help to determine the past. This problem will be at issue in the next but one section: 'Forward to the past'. But we may note at this stage that closed causal curves may be ruled out by an additional structural constraint on spacetime, namely the demand that spacetime be 'causally stable'.

5 *Stable causality*: whereas time-orientability provides a local sense to the direction of time, stable causality involves the assignment of a global time sense in a spacetime. This allows us to construct surfaces of simultaneity or 'time-slices' through every point of a causally stable spacetime. A given time-slice divides *all* spacetime into two categories: past and future – even though we should note that, in general, there is no unique way of making such divisions.[6] Hence, a curve passing through any given point to its future will remain in the future; and, since this future is globally defined, there is no way in which the curve can find its way into the past. This idea is one of the fundamental characteristics of spacetime theories. For we may determine both the differentiable and the conformal structures of a spacetime given the condition of stable causality.

Further constraints on causal structure may be added to these five fundamental elements. We shall see in the discussion of singularities in Chapter 10 that such additions may enhance the predictive power of spacetime theories.

BACK TO THE PAST

Backwards causation may be defined as a causal signal travelling between two events in spacetime such that the time direction of travel is from 'later' to 'earlier', i.e. the relation of causal precedence is opposite to that of temporal precedence. So backwards time travel may be described as travel from later events to earlier events. In a series of important articles on the subject, Michael Dummett explores the ideas involved in and the implications of backwards time travel and causation.[7] In the most recent of these, 'Causal loops', Dummett presents us with a number of stories concerning the appearances and disappearances of a 'time machine' and he notes that we might find it hard to explain exactly what is happening without invoking the idea of travel backwards through time. But he also points out that there is a cost: we may not be able to find an explanation for *every* phenomenon associated with a situation in which backwards causation is at work. Here, we shall examine three stories based on Dummett's account in 'Causal loops'. Dummett asks us to imagine something like the following situation:

A at 6 p.m. on Friday a small 'time machine' is placed on the left-hand side of an empty table top in a study; it has a clock attached which reads zero; the machine is activated and immediately the machine disappears;
B at 6 p.m. on Thursday (i.e. 24 hours earlier) the machine 'materialises' on the left-hand side of the table in the study with its clock reading zero;
C at 10 a.m. on Friday someone enters the room and moves the machine from the left- to the right-hand side of the table;
D the person activating the machine on the left-hand side of the table notices that there the machine is on the right-hand side of the table in the room; this clock stays in position and shortly after 6 p.m. its clock reads just over 24 hours.

The causal sequence here seems to be ABCD: the initial cause of the chain of events *seems* to be the activation and disappearance of the machine (A), followed by – its appearance at an earlier time (B) – its removal to another part of the table (C) – and someone seeing the machine in this new location (D). But, quite obviously, this is not identical with the temporal sequence BCAD: appearance (B) – relocation (C) – disappearance (A) – seeing machine in new location (D); see Figure 26 (p. 158).

How might we explain the sudden appearance of the machine at 6 p.m. on Thursday *without* invoking backwards time travel? Clearly, we would have to change our minds to some extent about the physics governing our world – especially if such things happen frequently. However, it seems that backwards causation commits us to some strange beliefs about the events in such a sequence. The fact that the machine is placed on an *empty* table top seems to require that, at some time between its appearance on Thursday at 6 p.m. and the machine being brought into the study 24 hours later, *the machine must shift its position*! Does this mean that whoever comes into the room on Friday morning is *compelled* to move the machine? If A, B, and C in the story above are to hold, then something must happen to relocate the machine. But what if we

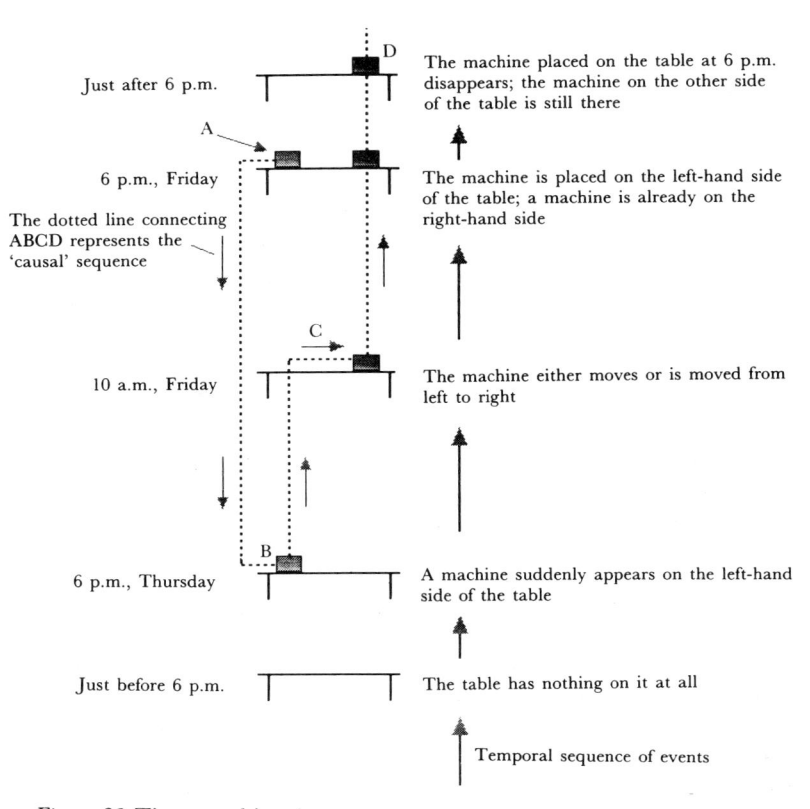

Figure 26 Time machine 1

were to make a determined effort to prevent anyone moving the machine – for example, we could lock the study for the 24 hours before the machine is activated. Now no one can get in and move the machine. There are two possible outcomes to such a manoeuvre:

1 despite our determination, either someone is able to get into the room and move the machine or the machine itself changes its position in some way; or
2 our efforts succeed, and the machine stays in its position on the left side of the table.

However, this second possibility clearly requires that the machine should be on the left-hand side of the table top at 6 p.m. on Friday.

If the machine is not there, then, *whatever* we do, we cannot succeed in any attempt to keep the machine on the left-hand side of the table.[8] If the machine *is* on the left, then a different story must be told which is consistent with this constraint, for example:

A a time machine, with a clock attached which is set at zero, is brought into the room just before 6 p.m. on Friday; on the left-hand side of the table there is another machine which is the same in all respects except that its clock reads almost 24 hours;
B at 6 p.m. precisely the machine with the clock reading zero is activated: on activation its clock begins and it appears to merge with the machine on the table;
C at 6 p.m. on Thursday a machine with a clock reading zero suddenly appears on the left-hand side of the empty table top – it remains there for 24 hours with its clock running normally;
D after 6 p.m. on Friday this machine stays in place on the left-hand side of the table top with its clock running on normally beyond 24 hours.

The temporal sequence here is CABD; the causal sequence is ABCD – bringing the machine into the room and activating it *seems* to be the cause of an earlier event; see Figure 27 (p. 160). Although there might seem to be two machines in this story of backwards travel, there is in fact only one. The machine brought into the room travels backwards in time, merges with itself, appears on the table, and remains there for at least 24 hours. The machine must merge with itself, or we cannot explain why the machine is found on the table when entering the room. This second story demands acceptance of two strange facts: first, assuming that the machine had been constructed some time before, an object is 'bilocated' for 24 hours – i.e. one and the same object is in two places at the same time; and, secondly, an object may merge with itself.

Whichever story we choose, we seem to be committed to accepting some rather odd beliefs about the world – and if we embrace the idea of backwards causation we have no immediate reason to object to either story. So we must accept that certain past actions or events are necessitated by future actions or events, even when these seem incidental or accidental – but this, of course, might seem strange only inasmuch as we are not used to thinking in terms of backwards causation: someone might think they have the option of moving or not moving the machine from the table, but in reality they have no choice at all. Hence, our freedom to act

Just after 6 p.m.		There is a machine on the left of the table
6 p.m., Friday		A machine is brought into the room, on activation it merges with the machine already on the left of the table
10 a.m., Friday		The machine is on the left of the table; it proves impossible to move it from that position
6 p.m., Thursday		A machine appears on the left of the table
Just before 6 p.m.		There is nothing on the table

The dotted line connecting ABCD represents the 'causal' sequence of events

Temporal sequence

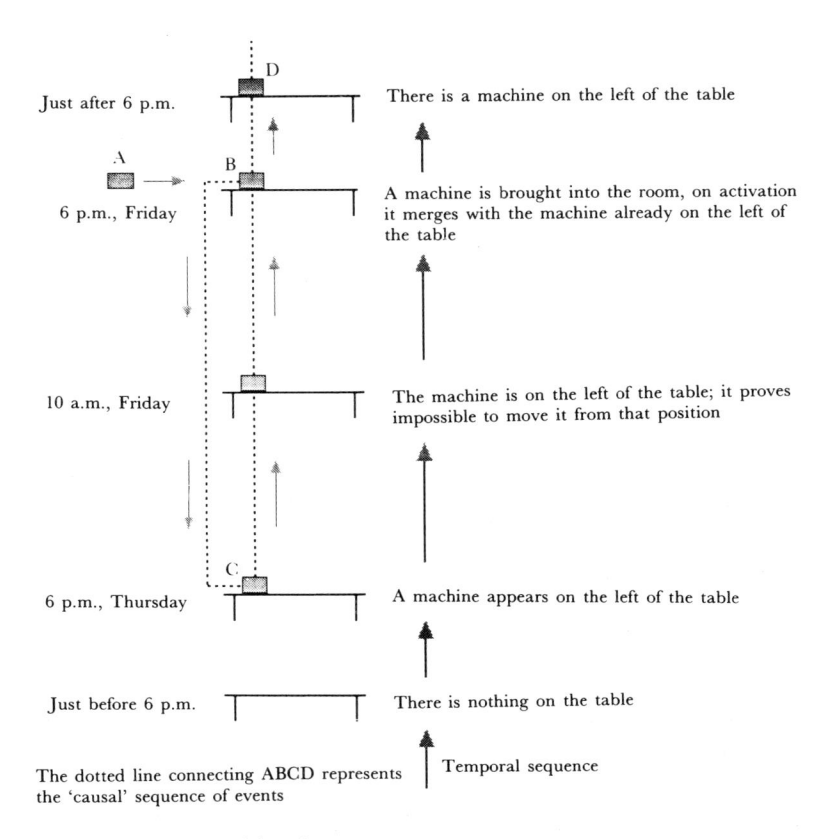

Figure 27 Time machine 2

may be constrained by future events. And we must accept the physical possibilities of bilocation and merging. We noted above that we might have to change our physics to account for sudden appearances of objects without invoking backwards causation. But it now seems that we shall have to change our physics in any case.

In each of the two stories so far, we have considered what seems to be an 'instantaneous' leap backwards by the time machine. Such a 'journey' does not seem to follow any kind of spacetime path at all. Hence these 'journeys' seem to be inconsistent with our ideas of travel in spacetime, which requires some sort of smooth passage along a spacetime path. It might be that the 'instantaneous' journey takes place via a topological anomaly – perhaps a wormhole in spacetime: although the two locations in spacetime do not seem to be adjacent, the topological features of spacetime may

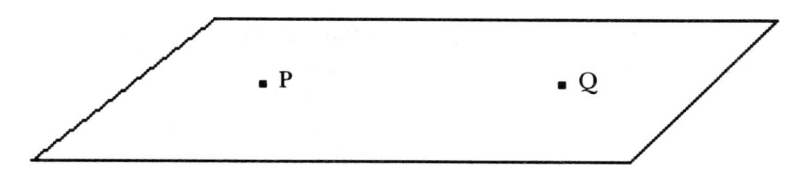

On a topologically flat plane, there is no short-cut from P to Q but if the plane is folded over so that P lies 'above' Q and the 'higher' and 'lower' parts of the plane make contact at P and Q, then a move may be made from P to Q without the necessity for a long journey along normal spacetime paths:

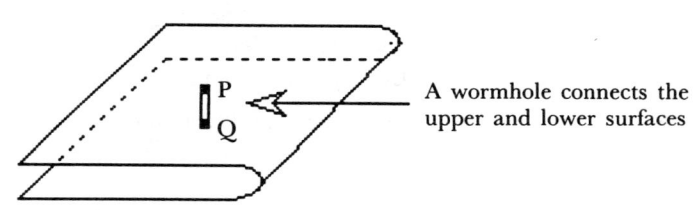

A wormhole connects the upper and lower surfaces

Figure 28 Idea of wormhole in spacetime

be such as to bring the two locations together via a wormhole; see Figure 28 above for an illustration of this idea.[9]

If, however, we imagine a time machine which travels more smoothly back in time (along a possible spacetime path, but in the past-directed direction), further complications arise. Imagine a machine which travels steadily backwards in time for 24 hours, and then travels forwards in time. We might characterise such as situation as follows:

A a machine M_2 is in position on the right-hand side of the table top just before 6 p.m. on Friday – it has a clock attached which reads almost 48 hours; on the left-hand side of the table there is another machine with a clock reading just a few seconds from zero – counting down towards zero so that at 6 p.m. it will read zero;

B a further machine M_1 with a clock set at zero is brought into the room, held above the machine on the left counting down to zero, and, at 6 p.m., M_1's clock is activated and it is dropped on to and merges with the machine below – M_1 is to travel backwards in time; as the machines merge at 6 p.m., they disappear leaving just M_2 on the right-hand side of the table;

C at any time in the 24 hours before 6 p.m. on Friday, two machines may be seen on opposite sides of the table: M_2's clock runs forwards from 24 to 48 hours, but M_1's clock runs backwards in time (i.e. this clock will appear to any normal observer to be running backwards from 24 to zero);

D before 6 p.m. on Thursday the table top is empty; at 6 p.m. precisely a single machine in the centre of the table with a clock reading 24 hours appears to split into two identical machines on opposite sides of the table – the only difference between the machines seems to be that one is running forwards and the other backwards;

E after 6 p.m. on Friday M_2 remains in place on the table top with the clock running on normally beyond 24 hours.

Once we accept backwards time travel and the possibilities of merging and bilocation, the only additional 'strange' occurrence in this story is the sudden appearance of a time machine which spontaneously splits into two nearly identical machines; see Figure 29 (p. 163). Why does backwards travel involve such a split? The story concerns just one machine: M_1 is the machine when it is running backwards in time; M_2 is the machine running forwards in time. The other machine on the table just before 6 p.m. on Friday is in fact M_1 travelling backwards in time. At 6 p.m. on Friday, the machine is placed as M_1 on one side of the table – by merging it with the machine seen in that position on the table; but, since M_2 is on the other side of the table at 6 p.m., it is obvious that at some time before then the machine *must* shift position. The splitting process is in fact the machine which has been running backwards moving across the table to start its 'journey' forwards in time.

If we enter the room during the 24 hours before 6 p.m. on Friday, then we will find that a number of actions, which we might attempt, turn out to be impossible. For example, if on entering the room we see *two* machines on the table, we could not destroy the machine travelling backwards no matter how hard we try. Even if we programmed a robot to enter the room and destroy the machine with the backwards-running clock if and and only if it detects two machines on the table, the robot too could not succeed in this task! For to destroy the machine with the clock running backwards implies that the machine travelling forwards would not be present – from the machine's 'point of view' the backwards phase is just an earlier part of its history. Some courses of action would be possible.

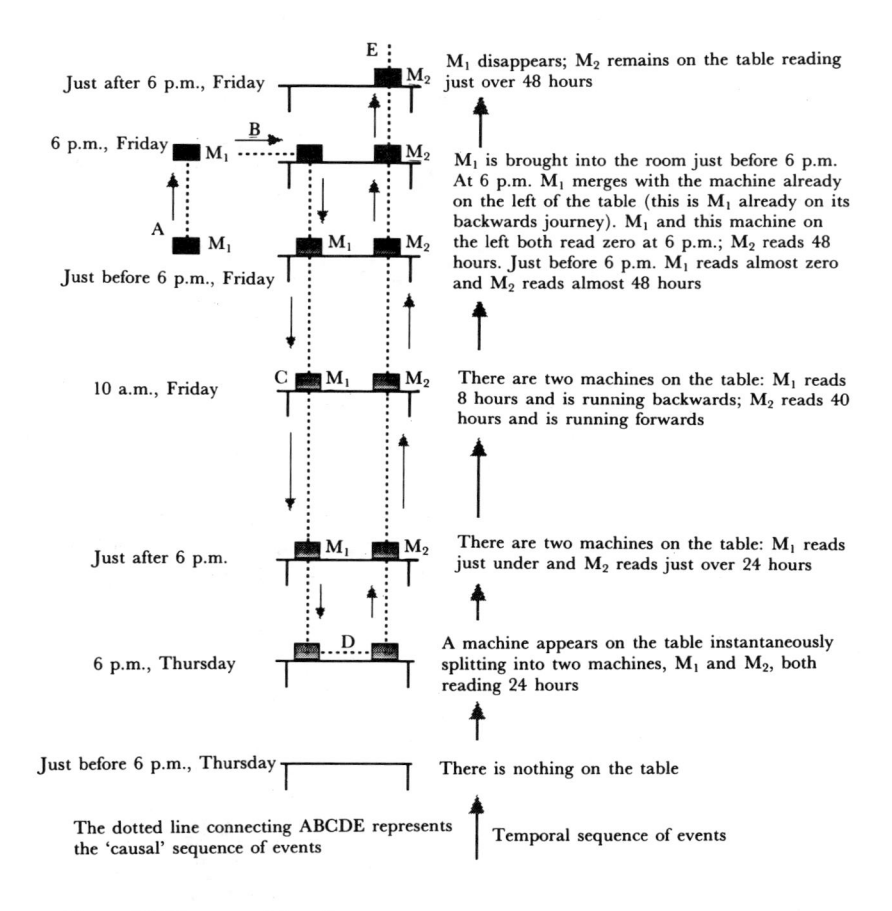

Figure 29 Time machine 3

We could move the machine travelling backwards – the spontaneous 'splitting' would then take into account the new location. If we destroy or move the machine from the table in its forwards phase, then there would simply be no machine already on the table when the time machine is activated at 6 p.m. on Friday.

There are many such stories which we might tell about travel back to the past. To accept them as physically possible means that we must be prepared to change our beliefs about both the physics in our world and our freedom to act. However, we might be inclined to regard all such stories as merely fictional entertainment.

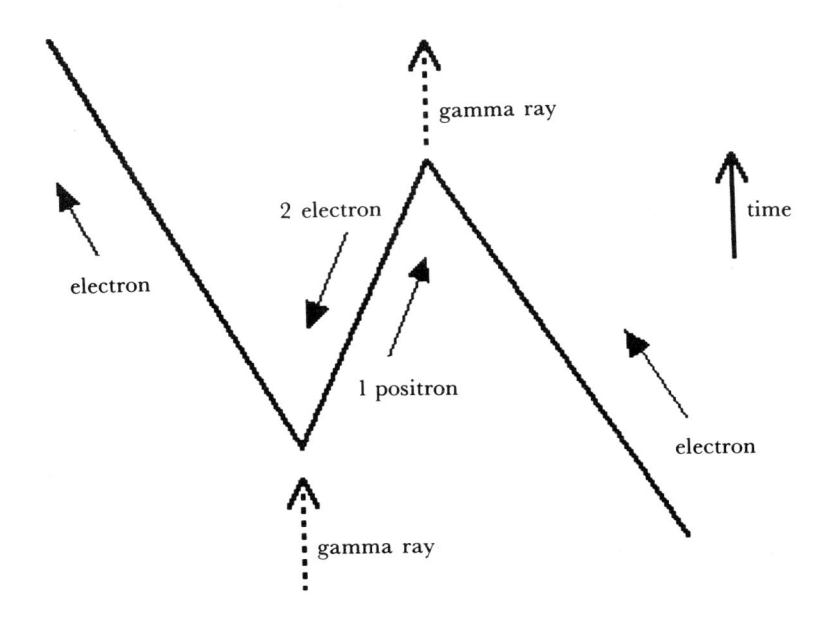

Figure 30 Positrons and electrons

Indeed, we might be entirely justified in doing just this – if there were no empirical basis whatsoever for backwards time travel. However, some physicists have suggested that certain physical events are far more elegantly explained if we use the idea of backwards time travel. Richard Feynman's 'positron theory' is one such account.[10] We may tell the story illustrated in Figure 30 above in two ways.

1. A gamma ray creates an electron and its antiparticle, a positron; the positron eventually meets another electron and both of these are annihilated resulting in the emission of a second gamma ray.
2. An electron in normal motion emits a gamma ray and immediately starts to move backwards in time until it meets with and absorbs another gamma ray when the electron once again moves forward in time.

The positron in the first account moves forwards in time; but Feynman characterises the positron as an electron moving backwards in time. What advantages does this alternative description have? Certainly, there is a sense in which Feynman's account is simpler than the orthodox version, since only one kind of particle is required: an antiparticle is simply a conventional particle moving backwards in time. But there is also a lingering suspicion that Feynman's alternative description is no more than a needless redescription. For the economy gained when we reduce the number of particles is counter-balanced by the loss in economy which results from our need to think in terms of backwards as well as forwards time travel. So, unless we are presented with independent evidence for backwards time travel, we might be inclined to stick with the orthodox version.

There are other possible instances of backwards time travel, but, rather than consider each possibility in turn, we need to look at Hugh Mellor's argument against both backwards travel and time travel in general.[11] Mellor maintains that it is always better to hold on to our normal account of time and causation and to accept, if necessary, the need for revisions to our physical theories. This argument turns in part on the claim that all stories about backwards time travel involve closed timelike or null loops around which causal signals may travel, at least in principle. Clearly, the third time machine story above does involve such a loop, for the machine completes a 'round-trip in time'. But, even if a machine travels backwards and does not return to the present, it will meet other objects which in principle could find their way to the machine's original location in space and time, thus providing a continuous closed loop. The first and second stories may involve a 'discontinuity' in time – we cannot trace any continuous 'path' for the machines backwards in time – they may just disappear from the later time and reappear at the earlier time, unless the machines in these stories travel via something like a wormhole in spacetime. But, even without wormholes, we might still trace a spacetime path directed backwards in time which connects the later and earlier spacetime locations of the machines. So, in a sense, we may still think in terms of closed loops in time even in such cases – providing we have reason to believe that past-directed travel is involved.[12] However, there are other, perhaps more plausible, ways of forming closed loops in time – by travel *forwards* to the past. So we need to discuss this idea before turning to Mellor's argument.

FORWARD TO THE PAST

A number of spacetimes in GTR seem to allow the possibility of a particle which follows a continuous future-directed path from a given starting-point eventually finding its way to other points which are located in the causal past of the starting-point. And, from the causal past of the starting-point, the particle may then reach the starting-point itself. Hence, we may construct closed non-spacelike (i.e. timelike or null) paths along which causal signals may travel in such spacetimes. As we have observed, even though the ideas of temporal and causal precedence may be inviolate in these spacetimes, the possibility of closed non-spacelike curves violates the condition of stable causality.

The simplest way to envisage the possibility of travel forwards to the past via closed loops in time is in the cylinder world, as described in Chapter 4.[13] Although a particle released from a 'starting-point' in such universes might always travel along future-directed curves, there always remains the possibility that it might return to its immediate causal past and perhaps collide with itself at the starting-point; see Figure 31 below. We should remember that a particle making such a 'round-trip' is not 'moving through time' in the way that we typically think of objects moving in space. The round-trip in spacetime is characterised by the *world line* of the particle. A particle does not move along a world line: rather the world line represents the particle's motion in spacetime. The world line describes the entire history of the particle around the closed path: it represents the particle's path in spacetime such that each point on the world line represents the spatial location of the

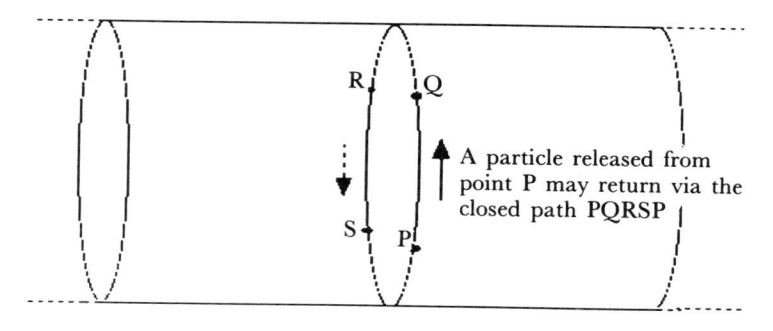

Figure 31 Closed timelike loop

particle at a different proper time (i.e. time as measured from the particle's point of view). To focus on a given point of the path is to find the particle there at a specific proper time. So it is in this sense that the particle is 'always' present at the starting-point as it is present at every point of its world line. Hence, a particle making such a 'round-trip' really could meet itself! But this meeting is a unique event. There is no question of the event happening over and over again. For each event upon a world line is a unique event in the history of a particle following the world line. And, if a particle could follow a closed path in time, then why not a specially designed machine with or without a person inside it?

This possibility allows us to construct all sorts of puzzling situations which some argue give us sufficient reason to dismiss closed non-spacelike loops as absurd. For example, we may imagine the following situation.[14]

First we make two background assumptions:

1 Two points P and Q lie on a closed causal loop: ... P ... Q ... R ... S ...... P etc.
2 The temporal order of these points is PQ, i.e. a future-directed curve from P may reach Q, and a future-directed curve from Q may also reach P by tracking around the closed loop in a single time direction.

Then we characterise a 'possible' situation on the basis of these two assumptions:

3 A device located at Q has two possible states: 'on' or 'off'.
4 This device at Q is 'on' if and only if a particle beam from P activates it and switches it on; if no beam activates it then it remains off.
5 The device at Q is 'on'.
6 When the device at Q is 'on', a particle beam is emitted which travels (via R, S, ... and so on) towards a second device at P; when the device at Q is 'off' no beam is emitted.
7 Therefore, given 3, 4, 5, and 6, a beam from Q is emitted in the forward time direction towards P – clearly this is possible only if the background assumptions 1 and 2 also hold.
8 The device at P has two possible states: 'on' or 'off'.
9 This device at P is 'off' if and only if a particle beam from Q switches it off; but, if no beam arrives from Q, then the device remains 'on'.

10 When the device at P is 'on', a particle beam is emitted which travels to the device at Q; when the device at P is 'off', no beam is emitted.
11 Given 7, 8, and 9, the device at P is 'off'.
12 Therefore, given 10 and 11, the device at P does not emit a beam.
13 Consequently, given 3, 4, and 12, the device at Q is 'off'.

Clearly, 13 contradicts 5. And making the assumption that Q is 'off' at step 5 does not help. For we shall then be forced to say that the device at P is 'on', emits a beam, and activates Q, switching it 'on'. Whatever we say about Q results in an outright contradiction. Without the two background assumptions which permit closed loops, then we obviously could not characterise any such situation – since it depends for its contradictory force upon step 7, which itself depends upon 1 and 2. Hence, the root of the contradiction seems to be just these assumptions. For many this kind of argument is a *reductio ad absurdum* of the idea of closed causal loops and therefore of any kind of travel to the past, given the fact that travel backwards in time also allows such loops to be constructed.[15]

Sometimes puzzles rather than paradoxes may be generated: we might return to the past with a portfolio of colour photographs of all Van Gogh's work and leave it with him when he is still a child; he then spends the rest of his life carefully copying the photographs![16] Sometimes the problems involving people may lead to paradox: if we can travel to the past, then we could kill our parents before they conceive us! What follows is based on one such story told by Jonathan Harrison.[17] Dum wakes up one day inside a large metal cabinet, remembering little about what has happened to him, but he does remember that his name is Dum. He notices a book close by with the title *How to Build a Time Machine*. He leaves the cabinet and pushes it into a deep river. He takes the book with him. He meets and marries a woman called Jocasta. They give birth to a son whom they name Dee. When Dee is 21, he finds Dum's book and, following the instructions, builds a time machine – it looks like a large metal cabinet from the outside. Dee persuades Dum to enter the machine. They set it in 'motion'. When Dee realises that the journey is going to be a long one, he kills and eats Dum to conserve supplies of food! Having eaten his father, Dee decides to change his own name to Dum. He then settles down to sleep. Dum wakes up one day inside a large metal cabinet remembering little about what has happened to him, but he does remember that his name is Dum . . . !

Does this story involve any outright contradictions? Certainly, we might begin to wonder how many people there are in the story, and where they came from. It looks as though Dum and Dee are the names used to refer to different periods of a single person's life: the person is born as Dee, murders someone called Dum whilst travelling to the past, changes his name to Dum, has a son called Dee, and is eventually murdered by his son. Clearly, Dee as Dum marries his own mother. But there seems to be no logical contradiction involved in the story. Some things in the story do not seem to have any definite physical origin. For example, the book is never written by anyone; it just exists and its history is described by the events on a closed loop in time. And although Dee is born in the normal way, we might also be puzzled about the origin of the genetic material passed on to Dee by his father Dum. There are no paternal grandfathers or grandmothers in this story. The single person is also bilocated as Dee and Dum for part of that person's history: strictly speaking the bilocation occurs at the beginning of Dee's life and the end of Dum's life; see Figure 32 (p. 170) which illustrates Dee/Dum's world line. However, these are physical not logical puzzles.

We could try to construct a contradiction by changing the story a little. For example, we might stipulate that Dee, remorseful after murdering his own father, commits suicide. If the suicide attempt is successful, then no one will emerge from the cabinet and no one will marry Jocasta and father Dee. So, if there is a suicide, there cannot be a suicide, for Dee will never be around to make the attempt. Hence, if Dee does commit suicide, then we do seem to have a paradoxical situation.

The existence of paradoxes in some cases is enough for some to say that time travel is impossible because it is logically impossible – we just cannot accept *any* story as possible because we can find some stories involving such contradictions. Others add to this the assertion that time travel would be effectively impossible in many causally unstable worlds – at least travel for any person, since far more energy would be required for such a journey than any technology we might imagine could actually deliver.[18]

If we wish to defend time travel against the accusation that it is impossible, then we might try to limit the possible situations which might arise from time travel to non-contradictory ones. In

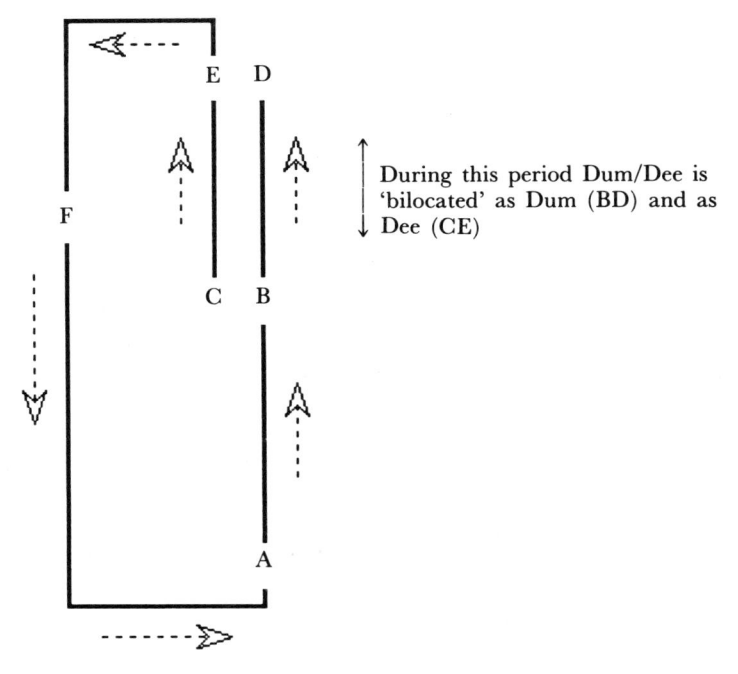

During this period Dum/Dee is 'bilocated' as Dum (BD) and as Dee (CE)

A: Dum awakes in time machine
B: Dum marries Jocasta and fathers Dee
C: Dee is born
D: Dum is killed
E: Dee changes his name to Dum
F: Dum is asleep as the machine travels back in time

If we look at this sequence from the point of view of a single person being born, living a life, then dying, we would trace the world line of Dee/Dum as CEFABD.

Figure 32 World line of the time traveller who killed himself

particular, we might try to block travel by people into their local past in order to attempt 'impossible' tasks such as murdering themselves as children, i.e. auto-infanticide! Paul Horwich, for example, argues that we can accept the possibility of time travel, but then we must acknowledge that certain situations – particularly those involving contradictions – will not be possible, given the observed behaviour of the physical world.[19] But this '*de facto*' argument might seem rather unsatisfactory: for we are not shown why contradictory situations cannot arise, we are simply told that they cannot because they do not. Nevertheless, that might be the way the world happens to be. In that case, it might seem churlish to complain that there are brute facts about the universe for which we have no convincing explanation.[20]

CORRELATIONS AND BACKWARDS CAUSATION

If an action carried out today can be a cause for an event which took place yesterday, then it seems natural to ask what might happen if we do not carry out the action. Could the event still have taken place? Yes, it could; but then some other action must have brought it about. But what if we can show a straightforward correlation between this kind of action and that kind of event such that the action is always some 24 hours later than the event. When the action is not carried out, we find no evidence of the event having taken place; and, when it is carried out, we always find that the event did indeed occur. For example, my action might be to make a wish (out loud) for a letter from my brother just before the postman arrives. But, for my wish to be fulfilled, my brother must have posted his letter the day before (at least!). I find that, whenever I make this wish out loud, there is always a letter from my brother in the postman's bag; however, if I fail to make the wish out loud, then there is never any letter from my brother. Must I conclude that it is my making a wish out loud today which causes my brother to post the letter yesterday?

I might decide that it is merely a coincidence. But, in this case, I would have to say most pairs of events which I regard as causally connected are merely coincidences – for all we have in such cases are the invariable correlations between the paired events. Whatever we think about correlation and causal connections, we should not apply one rule to forwards causation and one rule to backwards causation. The fact that correlations do hold is usually seen as an indication that correlated events are causally connected in some way. The stronger the correlation, the more likely we are to take the pair of events as causally connected – if not as cause–effect, then probably as the joint effects of some common cause. So we have no immediate reason to dismiss the correlation between my wish and the arrival of a letter as a pair of events which are not causally connected. And, if others find exactly the same kind of correlation between their making wishes out loud for a letter from their brother and the arrival of the letter wished for, then we might be inclined to accept that the correlation indicates that there is indeed a genuine causal connection.

One of the reasons why we might accept a correlation as a genuine causal link is on the basis of Mellor's assertion that 'causes make their effects more likely than, in the circumstances, they would otherwise have been'. Mellor goes on to explain how this is related to the ideas of correlation and causality:

> In order to discover causes and effects . . . science must at least establish statistical correlations between kinds of events which it claims to be causally related. The correlations must moreover be more than mere coincidence, i.e. they must have the force of statistical laws, however limited in scope. Otherwise nothing follows about how likely an event of the effect kind is to follow an event of the kind supposed to cause it. If hitting a window causes it to break, it must be more than a coincidence that windows of that sort more often break when hit like that than when not so hit.
>
> (Mellor 1981: 123)

So, for 'sensible decision-making' about causes and effects, Mellor argues that whatever else we might say about statistical relationships between events, we must invoke at least this notion that causes make their effects more likely. But now Mellor uses this low-key analysis of correlation and causality to show that the correlations between such events as my wishing for a letter and the prior posting of the letter should always be explained in terms of future-directed causation.

Imagine that we have a large group of people who have proved successful in the past in getting a letter from their brothers after wishing out loud for it. We decide to investigate the alleged causal connection between wishing out loud and receiving the desired letter. On a given day, say Monday, we find out which people in this group have brothers who have posted a letter to them that same day. Each letter posted has an electronic identification tag and miniature transmitter. This allows an automated computer system to keep track of each letter in the system; this system provides a continuous information print-out of the location of each posted letter. We then divide the group into two sections A and B based on Monday's information. If someone has been sent a letter by their brother, then they are placed in section A. And those to whom no such letter has been sent are placed in section B. We do not reveal to anyone in which section they have been placed. We then sub-divide the two sections. We ask half of the members in each section to make a wish out loud, and we ask the other half of the members in each section to do anything but make a wish out loud. Mellor suggests that in such situations the following outcomes are possible:

1 We find that the members of section A do exactly as instructed – half wishing out loud and half not wishing out loud; but now the strong previous correlation between wishing out loud and the arrival of the desired letter is lost – for half of the members of this section do not wish but still receive a letter; and we find that everyone in section B also carries out their instructions, so that a number of people wishing out loud do not receive a letter; hence, we would be far less inclined to say that the two events are causally related.
2 Everyone carries out their instructions and the strong correlation is maintained. Contrary to our expectations, formed on the basis of information collected about what was and was not posted, everyone wishing out loud receives a letter and everyone not doing so does not; somehow we were prevented from giving a correct account of what was and what was not posted – but this seems extremely strange, since much in the automated print-out would now be incorrect despite the fact that the system provided the print-out on the basis of the transmitted locations of each and every letter in the post.[21] And, as Mellor points out, our judgement that our account of a given event is reliable seems to depend upon the effects of that event rather than upon its causes: if we want to be sure that an earthquake has occurred in the San Francisco area, we do not rely on information about the causes of earthquakes but on the seismograph readings in that area, i.e. we look to the effects of earthquakes. Similarly, the automated print-out provides us with a reliable guide to whether or not a letter has been posted and to what then happened to a posted letter in the postal system. Why should all this evidence now be doubted? To accept this as a possible outcome would require us to question the reliability of all empirical evidence even when this evidence is as objective as might seem possible. We might also question the reliability of our evidence that the wishing ever

took place: why should we now accept any of the 'facts' about wishing as evidence for a correlation? Hence, we should be reluctant to accept this outcome as plausible in any way.

3 We find that everyone in section A, contrary to instructions, wishes out loud; and that everyone in section B, again contrary to instructions, does not wish out loud. Despite our efforts, the strong correlation is maintained. But, since none of the section members knows to which section they have been assigned, we cannot say that they acted on the basis of their knowledge about whether or not a letter had in fact been posted to them. This looks like a much more reasonable possibility, for here we need not revise our beliefs about what is to count as evidence. But we do need to account for the fact that the correlation is maintained only because the instructions are not universally obeyed – so we need to explain this disobedience. In fact, we might try to set up the situation so that disobedience seems out of the question. For example, all members of section A might be given a strong long-term sedative on Monday, 'guaranteed' to knock them until well after the postman has arrived on Tuesday with the letters from their brothers. For the strong correlation to be maintained now, we need more than disobedience: either our guaranteed sedative, which has worked on every occasion in the past, must fail to work in this case, or the members of the section must wish out loud in their sleep – even if we tape their mouths, they will rip the tape off in their sleep, breaking any other bonds to do this – or we must be prevented from administering the sedative. All of these possibilities might seem hard to explain. But still we can *imagine* that the correlation is nevertheless maintained by one or all of these strange occurrences taking place. Would we then be obliged to say that backwards causation is at work: that wishing out loud today can really cause someone to post a letter yesterday? Mellor says that we need not depart from our normal causal thinking, for everything in this possible outcome may be explained in terms of Monday's events being the cause of what happens on Tuesday. For example, we may say that the wish is made because the letter is posted, or that the sedative fails because the letter is posted, and so on. Of course, we should still be puzzled about the rather odd correlations in such a case. For example, why should my brother's action of posting a letter yesterday influence me today – is there some kind of 'telepathic' signal involved? But even with the backwards causal account we

are puzzled. For example, how is the wish transmitted backwards in time, why does this work with wishes about letters and not about other things, how does wishing today constrain the apparently independent activities of the postal service yesterday, and so on? In *both* cases we may need to revise both our physical theories and our general beliefs about the world.[22] If our account in terms of forwards causation seems problematic in any way, then at least we have the comfort of knowing that the account in terms of backwards causation is at least as unbelievable, if not more so.

Whether we are dealing with wishes and letters or the appearances and disappearances of time machines, we may always give an account in terms of forwards causation and there seem to be no serious disadvantages involved which are not shared by accounts in terms of backwards causation. Given that we may do this in principle, then we might be inclined to regard all the weird and wonderful inventions in this chapter as just that – inventions with no empirical basis. If the world did behave as oddly as the examples of time travel which we have encountered seem to demand, then we might think again about the *empirical* basis for backwards causation. But we still might question its *rational* basis and conclude that accounts in terms of forwards causation certainly leave us no worse and possibly rather better off.

9
EINSTEIN'S GREATEST MISTAKE?

INTRODUCTION

When confronted by the elegance and simplicity of a physical law, we might be forgiven for overlooking the presence of a constant in the equation before us. Sometimes a physical constant might be imposed upon a law after experimental investigations – as in the case of the Newtonian constant of gravitation G. However, sometimes the reason for the addition of a constant to fundamental equations might seem to be less convincing. Einstein's introduction of the cosmological constant into the field equations of the General Theory of Relativity (GTR) has often been criticised. The physicist de Sitter claimed in 1917 that the introduction of the constant into GTR 'is somewhat artificial, and detracts from the simplicity and elegance' of the original field equations, and recent writers, such as Abraham Pais and Stephen Hawking, emphasise the *ad hoc* character of the constant.[1] And Einstein himself is reported to have said that the constant was his greatest mistake – he had recommended getting rid of the constant in 1931.[2] In this chapter, we shall examine how the cosmological constant affected the development of GTR and assess the two criticisms above. I shall conclude that the introduction of the constant was neither *ad hoc* nor a blunder and furthermore that we can learn much about the status of fundamental laws from the history of the constant. We shall also find that the problem of the cosmological constant introduces us to another important area of debate within modern cosmology: does the universe obey a principle of anthropocentricity?

The cosmological constant has played a strange part in the history of modern gravitational physics. Welcomed into the fold of

GTR, then rejected, then rehabilitated, the role of this constant has important implications – philosophical as well as physical. The constant first appeared in GTR in a series of papers by Einstein in 1917 and 1918.[3] Einstein was then concerned with the global implications of his gravitational field equations – the laws governing motion in gravitational fields. In particular, he was troubled by four inter-related problems which had raised difficulties for many – from Newton himself to nineteenth-century physicists such as Ernst Mach, Carl Neumann, and Hans Seeliger:

1. Is space an infinitely extended three-dimensional container?
2. Does the universe have a finite matter content?
3. Is space itself the source of the inertial forces experienced by accelerating objects?
4. Are the 'fixed' stars really fixed – is the universe an essentially static place with the stars remaining roughly the same distances apart?

When Einstein introduced the cosmological constant, he believed that he had found a satisfactory solution to these problems within the context of GTR. Before examining Einstein's solution, we shall discuss not only Newton's own reactions to the issues raised by these problems but also the various responses of Mach, Neumann, and Seeliger.

SPACE AND INFINITY

In a letter to Richard Bentley in 1692, Newton explained his reasons for regarding the universe as an infinite arena containing matter everywhere:

> If the matter of our sun and planets and all the matter in the universe was evenly scattered throughout all the heavens, and every particle had an innate gravity towards all the rest, and the whole space throughout which this matter was scattered was but finite: the matter on the outside of this space would, by its gravity: tend toward all the matter on the inside and, by consequence, fall down to the middle of the whole space and there compose one great spherical mass. But if the matter was evenly diffused throughout an infinite space, it would never convene into one mass; but some of it [would] convene into one mass and some into another, so as to make an

infinite number of great masses scattered at great distances from one another throughout all that infinite space.
(Newton 1692)[4]

A finite amount of matter in an infinite space would effectively amount to a tiny cluster of stars in an immense void; and, as Newton realised, this cluster would tend to collapse in upon itself as a result of the gravitational attractions between the stars. The postulation of a finite spatial container for matter would not help. For the material in this universe would also gravitate inwards. There seemed to be only one straightforward resolution: an infinite number of stars must be spread reasonably evenly throughout space so that there would be no centre of attraction. Every star would be pulled equally in all directions; and so an equilibrium would be maintained. The only drawback to this solution – apart from the postulation of an infinite amount of matter – is the consequent lack of sense for the concepts of gravitational potential and force. Physics can only deal sensibly with finite quantities. Since there would be an infinite number of stars spread evenly throughout the universe, the pull on any given star would be infinite in all directions. However, this implies that the potential at any point and the force acting on a body at that point cannot be defined. For we can give whatever value we choose to the result of 'infinity minus infinity'.

This anomaly in the Newtonian perspective was tackled in the mid-1890s by Neumann and Seeliger.[5] In his book on Newtonian mechanics, Neumann remained faithful to the idea of an infinite space, but he saw a way to prevent the gravitational collapse of a cluster of a finite number of stars. He introduced a cosmological term into the equation for gravitational potential. Poisson's equation for the potential

$$\nabla^2 \phi = 4\pi G \rho$$

was changed to

$$\nabla^2 \phi - \lambda \phi = 4\pi G \rho$$

where λ is the cosmological constant. The effect of the cosmological term $-\lambda\phi$ is to balance the attractive force of gravity with a repulsive cosmological force. So the possibility of gravitational collapse is prevented by the existence of a force of repulsion acting over long distances.

The idea of a finite amount of matter in an infinite space presented a further problem. Such a universe approximates to a single system in a vast empty space. This is exactly the scenario assumed by Newton in his argument in the *Principles of Natural Philosophy* leading towards the postulation of space as an absolute entity.[6] As we have seen in Chapter 6, Ernst Mach, in *The Science of Mechanics*, responds dismissively to Newton's concept of space as an absolute frame of reference for motion: he goes right to the heart of Newton's argument and challenges the basic presumption that our laws might operate in any counterfactual circumstances. He argues in that book that we should refer motions to a material framework – that of the fixed stars. He sees no compelling reason for us to rely upon metaphysical artefacts like absolute space when all motions might be relative to a physical frame of reference. The inertial forces experienced by accelerating bodies might then be the result of some kind of interaction with the rest of the matter in the universe. However, Mach offers us no detailed positive explanation of how such a material interaction might operate; he stops short at his negative critique of Newton's views. But this critique involves an implicit attack on the idea of unrestricted abstraction. Although Mach employs the process of abstraction himself, he cannot bring himself to extend it to what he believes are absurd conditions. It is one thing to neglect frictional or electrical forces when making predictions, but he thinks it is quite another thing to neglect the entire contents of the universe. However, as we observed in Chapter 7, we need a clear indication of where we might draw the line between permissible and outlawed abstractions. Mach fails us, preferring polemical assertions to careful argument. So, at the beginning of the twentieth century, there was no clear reason for any physicist to dismiss Newton's argument for absolute space out of hand.

A final problem facing Einstein was that there seemed to be a rather far-fetched possibility presented by his field equations of gravity. They allowed for a dynamic universe – a universe in which spacetime itself might expand or contract; but all the astronomical evidence available in the early twentieth century suggested that we inhabit a static universe. Stars might live and die, but the structure of spacetime seemed to be unchanging – at least globally. Although Einstein had linked the structure of space and time to the distribution of matter and energy in the equations of GTR via the metrical notion of curvature, he saw no reason at all to depart from

Newton's belief that the nature of space as a whole is eternal and unchanging.

EINSTEIN'S UNIVERSE

Einstein, as he turned to cosmological issues, was therefore confronted with a number of beliefs which seemed to be consistent with Newtonian gravitation and with the available empirical evidence:

1 The universe is spatially infinite.
2 There is a finite matter content.
3 The matter content is globally static.
4 As Neumann demonstrated, a cosmological term may be introduced into the gravitational context.
5 The material universe approximates to a single system in an otherwise empty space.
6 Given Newton's argument in the *Principia*, the presence of inertial forces in accelerating bodies and systems must be explained in terms of motion relative to space itself.

Einstein's approach to these six statements was constrained not just by the empirical evidence but also by his long-term regard for Mach's anti-metaphysical philosophy of science.

The available empirical evidence suggested that the material content of the universe is both finite and static. And, once we accept statements 2 and 3, we are led directly to statement 4: a finite but globally static matter content would collapse in upon itself, so we need to find some reason why such a collapse does not seem to be happening. Hence, we may view the introduction of the cosmological term as a purely empirical matter; and this is so, regardless of whether the context is Newtonian or relativistic. Hence Einstein was inclined, like Neumann and Seeliger before him, to add a term – the cosmological constant – to his gravitational equations. As in the Newtonian case, the value given to the constant was such that it seems to act rather like a long-distance force of repulsion. Within the context of relativity, the addition of the constant produces a sourceless field which may counteract the 'attractive' gravitational field generated by the matter content. Set at the right value, the constant could produce the static global balance desired by Einstein.

But now Einstein was faced with a problem: to accept statement

1 as well as 2 leads us naturally to statement 5 and then to 6. Einstein was extremely reluctant to accept statement 6. To do so would be to embrace spacetime as an irreducible element in his account of gravitation. The equations of GTR allow the possibility of a finite amount of matter in an infinite space, and, as we have already noted, in such a universe space itself is an irreducible element in our descriptions of motion. It is one thing to explain gravity in geometrical terms and to demonstrate that spacetime is a dynamic structure which is influenced by the presence of matter and energy. It is quite another to admit that the structure of spacetime has a fundamental and irreducible status. For spacetime as a whole would be absolute in this sense: all inertial forces have as their ultimate source not matter but spacetime. This ran counter to what Einstein calls 'Mach's Principle' in his 1918 paper 'Principles of general relativity'.[7] This principle implies that any inertial forces experienced by accelerating bodies and systems must have their source in an interaction with the material contents of the universe. This is clearly in accord with Mach's conviction that matter and not space is the source of inertial forces. Einstein, who had read *The Science of Mechanics* as a young man, was much impressed by Mach's reluctance to embrace the concept of absolute space.

Hence, it was natural for Einstein to challenge statement 6 by attacking statement 1; and he did this by constructing a spacetime consistent with his equations, in which spacetime is spatially closed, i.e. in which space is finite. The 'Einstein universe' of 1917 is a spherically symmetric universe filled uniformly with matter – a homogeneous, vast, yet finite sphere. But the same problem which faced Newton still confronted Einstein: if the matter content is globally static, then this sphere will collapse. This propensity to collapse may be prevented if there is some repulsive force at work to counter the attractive gravitational effects. And this repulsion may be provided by the cosmological term. However, as we have seen, the question of collapse also arises in any static universe without an infinite amount of matter. It is not a problem which is peculiar to Einstein's static universe. If we do not assume that the matter content is static, then there is no immediate need for the cosmological term.

Although we might accept the general point, made by Pais and others, that Einstein's static universe was motivated by a Machian desire to repudiate absolute space, we should not accept that the introduction of the cosmological term as such into the gravitational physics of this universe was similarly inspired.[8] Four important factors need to be considered carefully here:

1 The paper of 1917 in which Einstein introduces his static universe begins by focusing on precisely the same problem as Neumann and Seeliger in their work of 1895–7.[9]
2 And, just as these two physicists were compelled to confront the problem of the extent of space in the Newtonian context, Einstein too could not escape this cosmological problem within GTR – and this would have been the case even if his basic intuitions had been anti-Machian. Hence, the 1917 paper discusses this problem in detail *before* moving on to his worries about the source of inertial forces.
3 We must distinguish carefully between the desire to establish Mach's Principle and the desire to incorporate the cosmological constant into GTR. Einstein needed to make a bold move to fulfil the first desire, since there was no direct empirical evidence at all for a finite space. But accepted observations backed up the second desire: the stars did indeed seem to be globally static and there seemed to be a finite number of them.
4 Because the cosmological constant represents a sourceless energy field, the constant itself is hardly a Machian artefact: Mach was strongly opposed to any putative non-material physical entity or phenomenon.

So the prime motivations for the introduction of the constant were:

1 the available empirical evidence; and
2 the need to tackle Newton's problem about the extent of space and its contents.

At worst, we might say that the constant is guilty only of association with Mach's Principle: for with the static universe not only did Einstein offer a resolution to Newton's problem but he *also* believed that he had found an economical way of realising Mach's dream of repudiating space as an irreducible element in our dynamical descriptions. However, it soon turned out that an anti-Machian universe incorporating the constant could be constructed: de Sitter found such a solution in which the source mass content is nil and therefore in which all motions must be referred to 'absolute' spacetime itself.[10] It is important to note that Einstein did not reject the idea of the constant immediately after de Sitter's anti-

Machian discovery. Had he done so, we might then have some reason to suppose that the motivation behind the constant was Machian. This gives further support to the argument above in favour of making a distinction between Einstein's Machian motivations and his broader empirical intuitions.

Nevertheless, the introduction of the static universe does highlight a certain tension amongst Einstein's commitments and beliefs. On the one hand, there are definite Machian influences – a high regard for empirical evidence and a strong desire for simplicity in our descriptions of the physical world.[11] But there is also a commitment to a 'constructive and speculative' and more intuitively theoretical approach not totally constrained by observational evidence.[12] As we have seen, the constant is clearly motivated by empirical considerations. But its introduction has immediate effects on the structure of GTR: there is a definite gain in generality with the field equations now being able to generate a much wider range of solutions – including Einstein's static universe; however, there is a corresponding loss in the mathematical beauty of the field equations – a simple, elegant picture of gravitation is replaced by a more complex view which adds a non-material energy field to the tools we need to describe the world.[13, 14] Furthermore, this field is clearly a theoretical artefact, lacking as it does any independent observational basis. This does not mean that the constant is an *ad hoc* addition, but it does illustrate that Einstein was willing to embrace theoretical entities despite Mach's influence.

An *ad hoc* change is most frequently taken to be a specific adjustment to save a theory in the face of some threatening (usually) observational evidence. As Popper points out, such changes invariably allow a theory to escape refutation.[15] However, GTR was in no real danger of refutation as far as the question of the global motions of the stars were concerned. Yes, Einstein did have an empirical motivation for adding the constant, but this was not contradictory evidence as such. The real problem was that GTR's equations seemed to lead to the unwelcome prediction of gravitational collapse. GTR fared no better than Newtonian theory in this respect, and it was natural for Einstein to offer the same remedy as Neumann and Seeliger. His reaction was to introduce a constant, the value of which would determine the large-scale dynamics of the universe. The result was a theory with increased scope which addressed a very real problem. Indeed both the need to resolve Newton's problem and the generality delivered by the addition of the constant provide persuasive evidence against the claim that Einstein's move was *ad hoc*.

THE COSMOLOGICAL CONSTANT: DID EINSTEIN BLUNDER?

Einstein may have considered the constant a mistake. But was its introduction really a blunder? It is frequently said that his preoccupation with the constant stopped him from exploring a much more rewarding answer to Newton's problem. Although the static universe possessed a high empirical pedigree, Einstein had shown that he was imaginative enough not to be tied to conventional wisdom. Had he taken a bolder approach, as writers like Pais maintain, then Einstein might have anticipated theoretically Hubble's celebrated observational discoveries in 1929 of the global recession of galaxies, thus paving the way for the hypothesis of an expanding dynamic universe. Yet, however much Einstein may have wished that he had not been so preoccupied with the idea of the static universe and that he might have had the foresight to predict an expanding universe, to cite the authority of Einstein's own retrospective antagonism to the constant is surely a good example of a superficially persuasive but illicit appeal! And he was surely not alone in missing the chance of anticipating Hubble. Newton might similarly be accused of a terrific blunder in not offering the possibility of an expanding material universe as a way out of his dilemma. If anything, the uncertain astronomical evidence as well as the general scientific climate of Newton's day might be said to be rather more supportive of the idea.

It might be said, against this, that the idea of a dynamic expanding universe is more natural within the context of Einstein's theory than within Newtonian gravity. For Einstein's concept of the metric of spacetime is essentially dynamic. However, the equations of GTR focus on the local dynamics of the metric of spacetime. Yes, we may build up a global dynamic picture – but we may also build a globally static universe. Without empirical cosmological evidence, there is no reason why any particular view should dominate. And so we must recall Einstein's commitment to empirical evidence. It would hardly have been in character for him to go *against* the then convincing observational evidence of a finite matter content and a static universe.[16] Of course, GTR makes some tremendous predictions. But we should note that the two major predictions made in the early days of GTR were related to empirical problems set firmly in the Newtonian context.[17] The calculations delivered by the field equations for Mercury's orbit and for the curved path of light close to the Sun were in both cases variations on a Newtonian theme:

1. The failure of Newton's equations to give an accurate account of Mercury's orbit had led to numerous attempts to save the theory by producing out of the theoretician's hat a host of plausible explanations, including Le Verrier's postulation of an extra planet close to the Sun.
2. Newton's theory together with a corpuscular theory of light certainly permits the prediction of light bending. Einstein himself had shown in 1911 – well before the formulation of GTR – that light rays must deflect in a gravitational field given the constancy of the velocity of light and the principle of equivalence. It is interesting to note that Einstein's 1911 prediction for the deflection of a light ray close to the Sun is numerically the same as that predicted by Newtonian theory. Only later did Einstein come to realise that the curvature of spacetime would affect the amount of deflection, and by the end of 1916 Einstein had calculated a deflection twice that of his earlier prediction.[18]

Hence, there is every reason to suppose that Einstein's imagination was constrained by his perception of the contemporary physical evidence. And to act entirely within character is hardly a mistake when that character had produced such extraordinary advances.

We might also argue that the addition of the constant was inspired in the sense that it set the climate for further research which was important not only for the development of 'classical' GTR but also for the difficult problem of understanding gravity in the extreme conditions close to singularities. Of course, we should not dismiss the fact that current observations set an upper limit of about 10^{-32} eV on the value of the constant. Even measurements designed to confirm that the photon has zero mass only put an upper limit on that mass of 10^{-16} eV. In fact, the cosmological constant is known to be closer to zero than any other physical quantity. Yet, although we may appeal to gauge invariance to set the photon mass at zero, there is no similar theoretical reason for us to set the constant at zero.[19] Furthermore, consider the following five points which indicate the important role played by the constant:

1 The constant undoubtedly increases the generality of the field equations of GTR – by adjusting the value of the constant, we are able to produce static, expanding, and contracting solutions. Given the uncertainty of astronomical evidence, especially at the time Einstein incorporated the constant, this must be seen as a definite plus. We might note that the paper in which Einstein introduces the constant does not specify any particular value for the constant; he merely demands that it be 'sufficiently small' to be 'compatible with the facts of experience derived from the solar system'.[20]

2 The mathematician, Alexander Friedmann, capitalised on this flexibility when he developed the 'Friedmann' models of 1922–4 which give theoretical rigour to these various cosmological possibilities.[21] These models, welcomed by Einstein in a brief note in 1923, have been of vital importance in developing our view of the universe's origins and global structure.[22]

3 Despite Einstein's dismissal of the constant in 1931 and the lack of any direct observational evidence for a non-zero value for the constant, it simply refused to disappear and like a 'mischievous genie' continually returned to haunt cosmologists.[23] The constant appears in numerous theoretically important solutions of GTR, such as Gödel's solution, which raises the possibility of travel to the past via closed causal loops, and the Oszvàth and Schücking model, as well as in the earlier Friedmann models.[24]

4 The constant has also been used to help physicists investigate possible links between GTR and electromagnetism; see, for example, Weyl's paper of 1918 as well as much more recent work in quantum gravity.[25]

5 Indeed, many cosmologists have come to regard the constant as an indispensable element in their description of the universe at early times. In quantum field theories, a vacuum is defined as the lowest possible energy density; so physicists are able to contemplate a vacuum which is empty of regular 'source' particles but not of fields! And such fields will have precisely the same sourceless character as that produced by the cosmological constant.[26] It is even suggested, in the powerful 'inflationary' models of the universe, that the rapid, early expansion of the universe may have been produced from a vacuum energy involving a very high value for the constant.[27]

All of these factors indicate that the constant has played a far from negligible role in the development of both classical and quantum GTR, stimulating rather than hindering progress.

LAWS AND THEORETICAL CHANGE

I believe that the tendency to regard the cosmological constant as an *ad hoc* mistake derives from a misplaced reverence for fundamental laws. If we treat a fundamental law as the one element of a theory which is immune to revision, then any suggested change, no matter how well-motivated, will be regarded as *ad hoc*: the only changes to be tolerated are those which bring revolutionary new theories with radically new laws. However, most scientific theories undergo almost continual scrutiny. Scientists are inclined to review not only the way in which the theory may be applied but also the basic ideas of the theory. Their reviews are not necessarily the prelude to a crisis of confidence in the theory; more frequently they are attempts to extend the scope of the theory. Such extensions may be achieved either by modifying the laws themselves or by adding to the theoretical context new but related laws as supplements to the original equations. The case of the cosmological constant shows just how willing scientists are to do this; but there are many, many other possible illustrations of this; for example:

1 Newcomb's suggestion that we may only account for the evidence of planetary orbits by recognising that the inverse square relationship of gravitation is only approximate and that very slight corrections must be made;
2 the multiplicity of formulations of the equations of quantum theory;
3 the changes which electromagnetism required for sound microscopic descriptions of such phenomena as that of dispersion; and
4 the adjustments made to the laws of classical mechanics to allow them to deal with far more than mass points, rigid bodies, perfect fluids, and smooth motions.

Hence, there is nothing unusual or heretical in Einstein's decision to add the constant to his field equations, thereby changing the form and content of the original laws of 1915–16. Laws are not, after all, untouchables. They are set within a changing theoretical context and they themselves may be modified without doing irreparable damage to the theory itself. For theories are complex constructions of which laws are only a part – but, granted, an important part. GTR, for example, is an intricate, interlocking context including: fundamental laws, empirical and philosophical principles, derived theoretical and experimental equations, local and cosmological solutions, fundamental commitments to geometrical and other entities, models, analogies, and physical facts and constraints, together with many links to other physical theories; we shall explore this context in greater detail in the final chapter.[28]

If theoretical laws could, even in principle, deliver the truth about the physical world, we might take a less tolerant line on using a range of laws within the same theoretical context. For we would then have a vested interest in matching a preferred law to the real world. However, Nancy Cartwright argues that theoretical laws *cannot* be strictly true in any real circumstance since such laws invariably deal with abstract situations which cannot be found in any empirical context.[29] Such laws may be accurate within their idealised domains, but they may only serve as guides to what happens in realistic situations. Hawking and Ellis provide a useful summary of both the power and the limitations of the field equations of GTR:

> Because of the complexity of the field equations, one cannot find exact solutions except in spaces of rather high symmetry. Exact solutions are also idealized in that any region of spacetime is likely to contain many forms of matter, while one can obtain exact solutions only for rather simple matter content. Nevertheless, exact solutions give an idea of the qualitative features that can arise in general relativity, and so of the possible properties of realistic solutions of the field equations.
> (Hawking and Ellis 1973: 117)[30]

Hence, the field equations apply directly only to highly symmetrical and simplified abstract contexts. An adequate characterisation of any realistic situation will require a less abstract, less 'exact' approach, but one which will have a greater chance of *direct* applicability to such a situation. Real situations demand a more flexible approach, drawing on perhaps several theoretical perspectives to produce an experimental or theoretical concoction with which we may characterise the context under examination more precisely. Such manoeuvres are used daily in contexts from the problems of experimental particle physics to the theoretical behaviour of singularities. Hence, there is even less reason for us to stick rigidly to a single law or set of laws which some may narrowly view as *the* law or laws of a given theory. And there is all the more reason for us to exploit whatever advantages a broad and dynamic theoretical context might offer.

The cosmological constant enriches the theoretical context of GTR. The main drawback is the lack of direct observational evidence for a non-zero value for the constant in the post-'big bang' era. We may have strong reservations about the likelihood of the sourceless field implied by the constant, but assertions about the *ad hoc* character of the constant are neither accurate nor helpful. For they deflect us from giving full credit to a theoretical role which certainly justifies us in regarding the constant as a useful and, on current evidence, still a potentially fruitful addition to GTR.

THE ANTHROPIC PRINCIPLE

The fact that the observed present value for the cosmological constant is so close to zero and the fact that significant positive or negative values for the constant would result in universes not fit for human habitation have led some writers to suggest that the value of the constant is low simply because it *has to be low* for life to develop, and they go on to propose a principle of anthropocentricity.[31] Anthropic principles focus our attention on good cosmic behaviour – where 'good' is defined in terms of what is consistent with life in the universe. There are two main versions of the anthropic principle: strong and weak. Each principle asserts a link between the physics of the universe and the conditions which must obtain for life to be possible. Brandon Carter, whose ideas about anthropocentricity stimulated this latest round of debate about the role of people in the universe, characterises the Strong Anthropic Principle (SAP) as follows:

SAP 'The universe must be such as to admit the creation of observers within it at some stage' of its evolution.[32]

The rationale for SAP depends in part upon the implications of quantum mechanics for the status of the observer, in part on the dimensionality of spacetime, in part on the possibility of multiple worlds within an all-encompassing universe, and in part upon a number of 'coincidences'.[33] These coincidences turn on the values of five 'fundamental' parameters involved in the evolution of the physical universe, including the cosmological constant.[34] If the cosmological constant and the other parameters were not very close to their observed values, then the universe would have either recollapsed very early in its history or expanded so rapidly that galaxies could not have formed – in both cases life could not have evolved, for the evolution of life needs time, the availability of complex atoms, and readily accessible energy sources. So a fine-tuned cosmic balance is needed for life to be possible. Is it merely a coincidence that the cosmological constant and other important physical parameters have the values they do? Or may we seek an explanation in terms of life itself? Supporters of SAP attempt to seduce us with two appeals: first, we are asked to focus on this universe in a narrow way – as it appears to be according to our current observations; and then we are asked to dismiss the idea that observed conditions and values could be mere coincidences and that life is therefore an accidental feature of the universe. SAP is typically justified by *de facto* appeals to the actual universe and its structures: we only have one universe; if the actual conditions and values for physical constants were not close to the observed conditions and values, then life would not be possible; observers exist in this universe; it is hard to explain why the conditions and values are as they are without reference to the existence of observers; hence, the observer is an essential part of the universe as given.

Paul Davies claims that SAP is motivated by positivism because, he says, it boils down to the demand that 'only that which is perceived enjoys true reality' with the consequence that 'a universe which did not admit observers is meaningless'.[35] Although there is an element of positivism in the desire to characterise the world as it appears to be in observation, SAP takes a step which no genuine positivist could condone: physical *necessity* is invoked to explain the connection between life and the observed conditions; as John Earman has pointed out, SAP makes a move from 'is' to 'must', from how things are to how things must be.[36] Positivists typically remain content with the level of mere description. There can be no necessity which attaches to the universe or processes and events within the universe. To make claims about how things must be, which is the essence of SAP, is, for the positivist, to engage in meaningless metaphysical speculation. Hence, SAP seems to be motivated more by the desire to provide a firm metaphysical foundation for cosmic coincidence. Indeed, some supporters of SAP engage in what might seem to be rather exotic metaphysical speculation when they argue that (a) the universe consists in a collection of 'worlds' which together represent all physically possible universes; (b) the universe we inhabit is really one of many worlds within this larger universe; and (c) our own world must exist and therefore life must exist, because this one *must* be present amongst the collection of possible worlds. Although some cosmologists favour such a worlds-within-world scenario consistent with inflationary cosmology, there is little evidence to suggest that this view should be taken too seriously.[37]

For these reasons, SAP is a highly speculative idea to which few cosmologists would wish to subscribe. It carries with it perhaps too many egocentric overtones for most cosmologists: 'the universe exists so that we might exist' is a little too much to swallow, even if as a result we are left with one gigantic set of cosmic coincidences. Why should we have to explain every coincidence? And, more problematically, should we be satisfied with an 'explanation' which spells out the conditions in and development of the universe in terms of our existence – not by providing some straightforward mechanism linking our existence to the actual conditions and development, but by asserting the link as an 'irreducible' fact? As Paul Davies points out:

> From the strictly physical point of view it seems mysterious, to say the least, that the existence of conscious beings can actually bring about the celebrated coincidences. Clearly any direct causal connection is impossible. Special physical conditions may produce man, but man can hardly be attributed the credit for establishing his own environmental requirements.
>
> (Davies 1982: 122)

The weaker principle is less controversial and more widely accepted, for it acknowledges the fine balance required for life to exist, but does not seek to impose this balance on the universe as a matter of necessity. Stephen Hawking states the Weak Anthropic Principle as follows:

WAP 'Intelligent life can exist only in certain regions of a given universe with given physical laws.'[38]

This principle is based on Carter's assertion that 'what we can expect to observe must be restricted by the conditions necessary for our presence as observers'.[39] WAP does not force us to explain the laws and conditions in the universe in terms of life and consciousness; the explanation should be the other way round: life and consciousness depend upon specific conditions and laws holding at least in certain regions of spacetime. Hence, the fact that the observed value of the cosmological constant is what it is may have a central part in the explanation of why life is possible in this universe. Carter maintains that the essential function of WAP is to alert us to 'the risk of error in the interpretation of astronomical and cosmological information unless due account is taken of the biological restraints under which the information was acquired'.[40]

The cosmological constant is just one item in a list of potential problems for the well-behaved universe, some of which might adversely influence the possibilities for intelligent life and others which might simply interfere with our ability to make predictions about physical phenomena. Black holes, the big bang, and topological anomalies in spacetime all present problems for GTR. Although the focus of the next chapter will be on the beginning of the universe, we shall also review the causal problems associated with black holes and other 'singularities' in spacetime. In doing so, we shall obtain a far more detailed vision of the complexity and range of GTR's theoretical context.

10

COSMOLOGICAL CONUNDRUMS

INTRODUCTION

The Newtonian universe has a beginning in time for matter, but not for time itself. God is supposed to have created the material world, probably in stages, starting at some definite point in the past. Before matter appeared, there was a material void: an empty and eternal spatial arena. We might ask the Newtonian: what happened before creation? For the creation is a moment in a temporal framework which stretches infinitely to the past and to the future. God's eternal character is typically related to this infinite span for time. Space, time, and God all existed before the creation. God chose a particular moment to create the material universe; and God, like time and space, will presumably continue to exist after the end of the material universe. Without God, the appearance of matter at a given time might seem somewhat mysterious. But, of course, any reference to God as creator simply replaces one mystery by another.

Leibniz found much about this Newtonian view objectionable. The main difficulty highlighted by Leibniz is this: why did God create the material universe at one time rather than another? What possible reason could God have had for choosing one instant out of eternity over any other time? As in the similar problem, discussed in Chapter 5, about the indiscernibility of two universes which are distinct only in their respective spatial positions, Leibniz appeals to the Principle of Sufficient Reason:

> In things absolutely indifferent, there is no choice; and consequently no election, nor will; since choice must be founded on some reason, or principle. A mere will without

any motive, is a fiction, not only contrary to God's perfection, but also chimerical and contradictory. . . . Since God does nothing without reason, and no reason can be given why he did not create the world sooner; it will follow, either that he created nothing at all, or that he created the world before any assignable time, that is the world is eternal.

Leibniz (1716)[1]

So Leibniz argues that, if God creates matter at all, then the Newtonian must accept that there has always been matter and give up the belief that God created the material world in a pre-existing spatial arena. But this move presents obvious theological difficulties for someone who wishes to accept something like the free creator of Genesis: if matter is eternal, then not only is the creation story of Genesis explicitly attacked, but the existence of matter becomes a necessary feature of God's plan, rather than something which is contingent upon God's will. Leibniz believes that the Newtonian must therefore abandon his position. Clarke, in defence of the Newtonian view, appeals to God's eternal point of view and to his inscrutable design:

> It was no impossibility for God to make the world sooner or later than he did: nor is it at all impossible for him to destroy it sooner or later than it shall actually be destroyed. As to the notion of the world's eternity; they who suppose matter and space to be the same, must indeed suppose the world to be not only infinite and eternal, but necessarily so.[2] . . . But they who believe that God created matter in what quantity, and at what particular time, and in what particular spaces he pleased, are here under no difficulty. For the wisdom of God may have very good reasons for creating this world, at that particular time he did; and he may have made other kinds of things before this material world began, and may make other kinds of things after this world is destroyed.

Clarke (1716)[3]

So, Clarke asks, who are we to question how God acts and when he chooses to act? Leibniz believes that this response is inadequate and does not do justice to God's inherently perfect, rational will.

Rather than admit that God may have his own reasons for acting — reasons which are inaccessible to the finite human mind, Leibniz argues that space and time are mere appearances. We do not see reality as it is. We view reality as a spatial, temporal, and material domain. But this is not its real nature. Reality is the realm of monads: the basic metaphysical stuff which forms the basis of all creation. Spatial, temporal, and material properties are not essential features of monads themselves, they are features of the way *we experience* monads. Hence creation for Leibniz is not creation in time, it is creation of all that gives rise to experience, including time. From a human point of view, God acts to create all matter, space, and time 'all at once'. It is as if God sculpts some vast four-dimensional image, a total history of the universe, in a single divine 'act'. Clearly, this act cannot be the kind of action which you or I perform, since such actions always take place in time. The act of creation here involves the creation of time itself.

Does Leibniz's suggestion make sense? If we set aside the obscure metaphysics, we might identify the central idea more clearly: space, time, and matter come as a package deal. The fundamental link between matter and spacetime in GTR might provide some support for the essential idea here. Whatever we take to be creation is the creation of everything, including time itself. If we insist on thinking in terms of an actual creator, then that creator is transcendent or 'outside' time and space, i.e. has a character which is neither spatial nor temporal. So, from such a creator's point of view, it is more appropriate to think of the creation of all history in a single 'act' rather than of the creation of the material world in a spatial arena at a given time.

If we are to appreciate just how sensible these ideas might be within the context of modern cosmology, then we must review the claims made by cosmologists about the big bang – the initial singularity – sometimes said to be the beginning of time itself. And the problem of the big bang will then lead us to two related cosmological conundrums:

1 Where did the matter content of the universe come from? Should we accept the central claim of recent work in inflationary cosmology that its creation was literally from nothing?
2 Can singularities like black holes bring causal disruption and indeterminism to the rest of spacetime? Or are we allowed to advocate the cosmic censorship hypothesis and draw an impenetrable cloak around every singularity?

Although few cosmologists dare to offer final answers to such problems, by exploring their ideas, we may nevertheless obtain a

TIME, SPACE AND PHILOSOPHY

clearer view of the questions which we should be asking about the large-scale structure and history of spacetime. And we might also learn why some questions just do not make sense.

THE BIG BANG: A SINGULAR IDEA

The standard 'big bang' view of modern cosmology starts from three basic assumptions:

1 The Friedmann models of GTR are essentially correct at least in their large-scale characterisation of the universe. The universe is homogeneous and isotropic: the same everywhere and in every direction. As we noted in Chapter 3, this assumption has a high empirical pedigree.
2 Each cluster of galaxies seems to be moving away from every other cluster of galaxies: so that the universe is 'expanding' on a

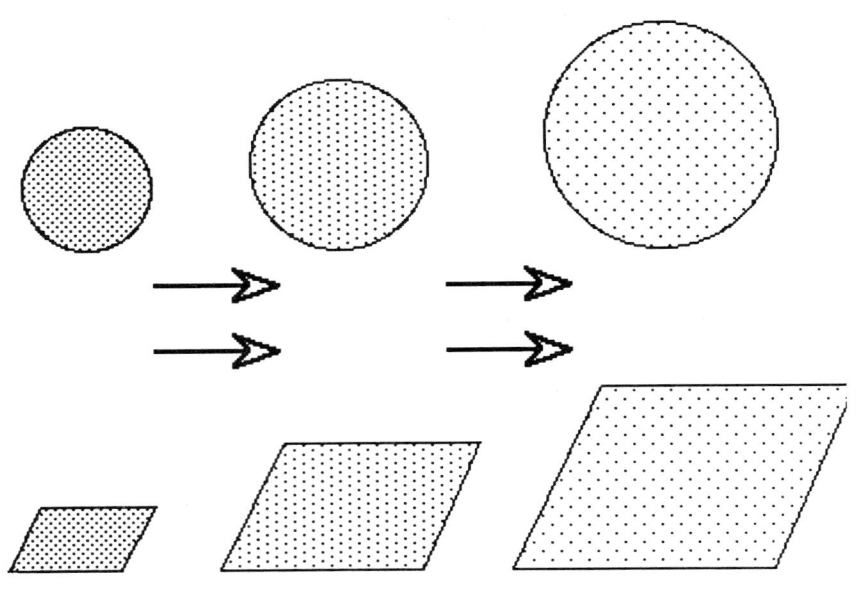

As spacetime expands, the material in the universe becomes less compressed. The two sequences above illustrate the expansion of a spatially closed spherical universe (rather like a balloon expanding) and the expansion of a spatially open plane (of which only part is shown here)

Figure 33 Expanding sphere and plane

grand scale. The field equations of GTR tie the geometry of spacetime to the distribution of matter. So, in the Friedmann models of the universe, the matter distribution is expanding because *spacetime is expanding*. Cosmologists frequently cite the analogy of an expanding balloon to illustrate the general idea involved: imagine a balloon with spots spread evenly (homogeneously) over its surface; as the balloon expands, the spots 'move' further apart; see Figure 33 (p. 196). The spots act like clusters of galaxies, and the balloon like the structure of spacetime. The expansion is of spacetime itself.

3 Clearly, if spacetime is expanding, then at some time in the past spacetime and matter must have been tremendously compressed. Matter in such a condensed form would have a rather different character compared with the matter around us now: even the simplest atoms would be broken down into their fundamental parts in a very hot and extremely dense 'radiation soup', far more active and violent than the hottest star. Evidence for this assumption was provided in 1965 by Arno Penzias and Robert Wilson, who found the 'microwave background': the cosmic remnants of this early, radiation-dominated phase of the universe; see Figure 34 (p. 198).[4]

So GTR, together with some very plausible empirical assumptions, leads us directly to the hypothesis that the matter (or radiation) content of the universe becomes more and more tightly compressed as we look further and further back into the past. The earliest stage of all may be characterised as the initial singularity: all material compressed into no space at all – a 'point' of infinite density and therefore with infinite curvature.[5] The expansion of material (and therefore spacetime) away from the initial singularity is rapid and explosive – hence the idea of a big bang.

A more precise definition may be given to the idea of a singularity using the concept of a geodesic. The world line of a freely moving object is called a geodesic, and the motion of the object may be described by its 'affine parameter'. If there are no singularities in space which might mark the 'end-points' of geodesics, then we would expect the affine parameter which describes the geodesic to assume all values from plus to minus infinity, thus signifying that the geodesic goes on 'for ever'. A geodesically incomplete spacetime is one in which at least one geodesic comes to a halt. The geodesic would reach an 'end-point'

An extremely brief history of the universe:

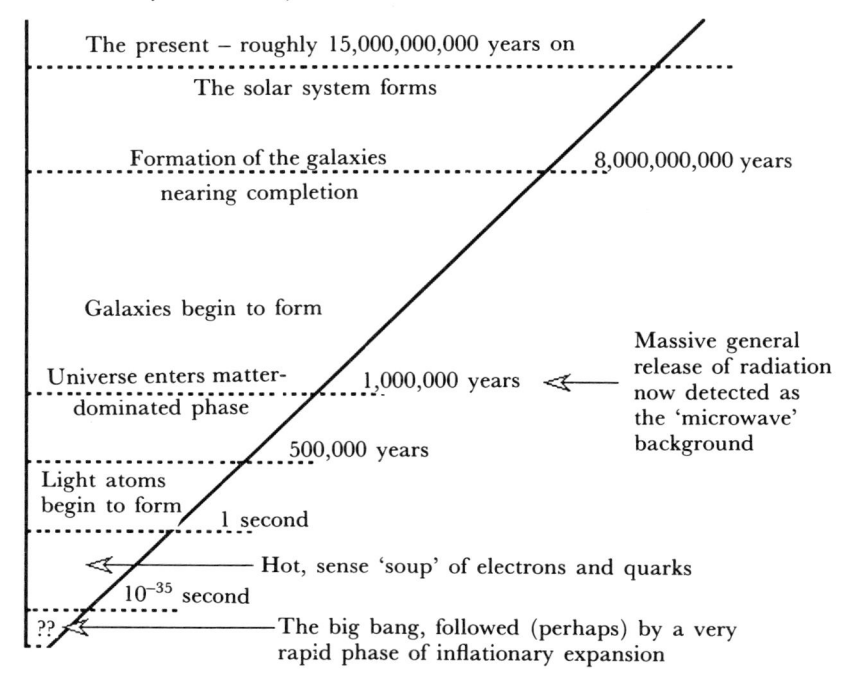

- The present – roughly 15,000,000,000 years on
- The solar system forms
- Formation of the galaxies nearing completion — 8,000,000,000 years
- Galaxies begin to form
- Universe enters matter-dominated phase — 1,000,000 years ← Massive general release of radiation now detected as the 'microwave' background
- 500,000 years
- Light atoms begin to form
- 1 second
- ← Hot, sense 'soup' of electrons and quarks
- 10^{-35} second
- ?? ← The big bang, followed (perhaps) by a very rapid phase of inflationary expansion

Note: not to scale!

Figure 34 Back to the big bang

if spacetime comes to an end. And singularities are 'places' where this can happen. We can draw an analogy between physical singularities and topological holes to make this idea clearer. If a spacetime were to have sets of points literally cut out from it producing a topological hole in spacetime, then any geodesic arriving at the excised region would have nowhere to go. The topological hole would be a terminus for some geodesics and a starting-point for others. A physical singularity behaves in essentially the same way as far as geodesics are concerned – the singularity provides a definite terminal boundary for geodesics; see Figure 35 (p. 199). Hence, we may define a singularity as a physical phenomenon which produces a geodesically incomplete spacetime.[6] Some singularities, such as black holes, may be

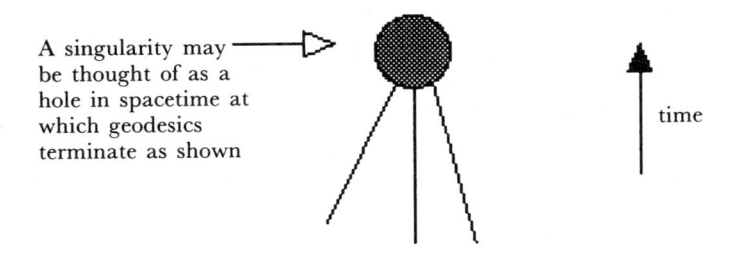

A singularity may be thought of as a hole in spacetime at which geodesics terminate as shown →

time ↑

Figure 35 Geodesics meeting a singularity

throughout spacetime. But the initial singularity is a fixed terminus for all spacetime.

THE BEGINNING OF TIME?

It might seem very tempting to ask at this point: what happened before the big bang? Three main responses have been given to this question:

1 The first (and now standard) response is to say that the question does not make sense! Since spacetime begins at the initial singularity, there is no sense in the idea of time before the singularity. For all time is by definition contained within a spacetime which is 'this side' of the singularity. It just does not make sense to talk about 'the other side' of the initial singularity, for our characterisation of spacetime is such that there simply is no other side, no earlier time. Similarly, if there is a final singularity, when all matter and spacetime collapses in a 'big crunch', there would be no sense in the idea of a time 'later' than the final singularity. If we accept the idea of an initial singularity, what we cannot say is that there was any earlier time. Similarly, the general idea of spacetime prevents us from giving any sensible answer to the question: what is spacetime contained in? For the idea of containment is itself spatial. We cannot invent higher *spatial* dimensions in which to locate spacetime. The only space we have is already given fully in spacetime. So all we have is spacetime, and there is no sense in the ideas of a time before or a space beyond.

2 However, we might argue there is no genuine initial singularity, that the rapid expansion of the universe follows immediately after a massive collapse, and that the universe 'bounces' from collapse to expansion when some minimum volume is reached. What happened before the big bang would then be an intelligible question. We might say, for example, that the present expansion is just one phase in a universe which has existed already for an infinite time, that the universe continually expands and collapses over and over *ad infinitum*. Or we might be more cautious and say that there were times earlier than the big bang, but that all we may say about the universe before the big bang is that it was in a state of collapse just before the big bang.
3 We might also accept the idea of overall expansion but deny that there was *any* initial highly compressed state. The steady-state theory, popular until the discovery of the microwave background, asked us to accept a universe in which matter is being created continuously throughout the universe, in a stable and uniform manner. The steady-state model developed by Bondi, Gold, and Hoyle was based in part on the static universe devised by de Sitter in 1917.[7] The expansion in the steady-state model is caused by a repulsive force linked with the cosmological constant. In this view too there is no first moment: the creation of matter has always happened and will always happen. But on this view there is no reason to suppose that there will be any widespread remnants of an early phase of the universe, dominated by radiation. Hence, it was dropped like a hot potato by most cosmologists as soon as Penzias and Wilson found the microwave background.

The first response, for all its audacity, is the most conservative of the three. For we have no firm evidence to suggest that the universe has ever 'bounced' from collapse to expansion. Indeed, it is hard to imagine what kind of observational evidence might indicate a bounce rather than a singularity. Nor is there any convincing empirical evidence for the idea of continuous creation despite the feverish, *ad hoc* adjustments made to the steady-state theory by its supporters in their attempts to make it more acceptable after the discovery of the microwave background.

Why should anyone wish to resist the idea of a universe evolving from an initial singularity? This may be because the idea of a universe which is infinite in time, past as well as future, has a pleasing symmetry. The standard cosmological view involves a universe expanding from an initial 'boundary' for ever. The future is indeed infinite. But the past is finite. Although there are suggestions that the universe could collapse in on itself, there is as yet no *strong* observational evidence to support this view, as we noted in Chapter 3. Hence the universe seems to be curiously asymmetrical. However, even if there is sufficient matter in the universe to induce a global collapse, the most likely result would be a universe which is finite in time, future as well as past. So, even with singularities, the universe could still exhibit a global symmetry in time.

Some might decide to favour the second and third responses because of the implicit refusal to contemplate a 'first' moment in time. The idea of a first moment in time carries with it, in some eyes, the suggestion of a creation. So an alternative view of the universe, involving no such first moment, might bring with it the advantage of no creation and so no creator. But this manoeuvre fails: for we can still imagine a four-dimensional universe being created as a package deal along the lines suggested by Leibniz. It may be an unnecessary extravagance to think of the universe as created by God. The world might simply exist and require no such justification.[8] But nothing prevents believers in God from adding the claim 'and all spacetime was created in a single divine act' to their other beliefs. So there seems little reason to resist the idea of a first moment just because of some general antagonism to the possibility of a divine creation.

A more plausible reason to prefer the second or third response might involve the recognition that GTR, as a classical theory of spacetime, just cannot deal with either the physical infinities which arise at singularities and the quantum effects which are likely to occur in their neighbourhood. Stephen Hawking describes the problem as follows: work on singularities in the 1960s and the early 1970s produced a number of singularity theorems which

> showed that if classical general relativity were correct, there would inevitably be a singularity at which all physical laws would break down. Thus classical cosmology predicts its own downfall. In order to determine how the classical evolution of the universe began one has to appeal to quantum cosmology and study the early quantum era.
>
> (Hawking 1987: 632)[9]

So some might be inclined to look for a cosmological account which does not involve singularities and therefore does not lead to the breakdown of classical relativity. But it seems absurd to exclude the initial singularity when other singularities, including black holes, are an essential element in our general account of the universe. Black holes, like all singularities, are likely to involve quantum effects in the hot, extremely dense environment which they produce. And, like the initial singularity, black holes are predicted by GTR. So, however problematic singularities might seem to be, we have no reason to try to avoid dealing with the quantum effects which are associated with them.

However, we might still be perplexed by the idea of a physical singularity within spacetime: for it implies that there exists an actual infinite quantity in a given location. Can we accept that cosmology might involve quantum effects, that the universe really is evolving from extremely hot, dense initial conditions, and that there is no previous 'pre-bounce' universe, but also say that there is no initial singularity as such *in* or at least on the edge of spacetime itself? Stephen Hawking and Jim Hartle suggest a way for us to do just that. They ask us to accept the idea of an initially highly compressed state and deny the possibility of an actual first moment.[10] Their explanation is based on the fact that spacetime need not be *closed* at the initial singularity: it may be *open* instead. The idea that a singularity provides a starting- (or finishing-) point for a geodesic requires us to consider the singularity as a physical entity of some sort. The initial singularity may then be thought of as providing a definite boundary to spacetime. But Hawking and Hartle suggest that there may be no boundary and no singularity: 'the boundary condition of the universe is that it has no boundary'.[11] Only if spacetime is closed at the initial singularity will there be a definite sense to the idea of a first time. If we pick out two definite 'end-points' for an interval on a metre rule, say 10 cm apart, then the interval between these points may be open or closed at one or both ends. If we include the end-points in the interval, then the interval would be closed at both ends. But, if we do not include the end-points, the interval would be open at both 'ends': there is no end-point at all but an infinite number of points each closer to the 'end' but never reaching the 'end' – and this is because there is literally no end; see Figure 35 (p. 199). If we track a geodesic back through time, then the geodesic may simply approach the so-called initial singularity and never reach it. It does not reach it because there is literally nothing to reach. The geodesic need not have a definite end-point at the singularity for in such an open-ended spacetime there is no definite boundary with a singularity. This idea allows us to consider the possibility of an open-ended spacetime which has no definite beginning.[12] However, it might be appropriate to recall some of the worries about the status of the continuum raised in Chapter 1. Why should we rely on the received wisdom about the mathematical properties of spacetime structure and allow the continuum model, which underwrites the account of open and closed intervals, to dominate our view of early cosmology? Even if we accept that the continuum model provides a fairly good approximation to the spacetime around us now, we may reserve our judgement about such early times.

Modern cosmology allows us to approach the debate between Leibniz and Clarke with a wider variety of options. If we insist on some sort of divine 'creator', the four-dimensional perspective of GTR does seem to support Leibniz's (non-metaphysical) ideas rather more than the Newtonian account. Clarke, like Newton, did not imagine that matter, space, and time might come as a package deal. The creation of matter was, for them, a mysterious event. And the existence of an (uncreated) eternal space, which has real physical properties in the Newtonian view, was also a mystery. GTR provides us with a way of characterising space, time, and matter such that the question of creation may be avoided if we choose to do so. Talk of what lies outside spacetime has no physical meaning; so talk of something not in spacetime creating spacetime has no *physical* meaning. Of course, it is up to you to decide whether there is a transcendent deity 'beyond' spacetime with some kind of non-physical character responsible in some sense for the physical domain. But that decision just cannot be anything other than metaphysical speculation, perhaps backed by some supposed non-physical 'experience' of God. However, modern cosmology, as a physical theory, must confine itself to the physical, to what lies within spacetime.

Yet it is hard to suppress the question: where did all the material in the universe come from? But there is no need to answer once we accept that spacetime and its contents might simply exist as a four-dimensional entity, with or without boundaries. The presence of matter would then be just a characteristic of this four-dimensional universe. However, recent work in inflationary cosmology indicates that we might provide an answer which further undermines the need for a metaphysical creator to explain the presence of the physical. Where did all the matter come from? The answer might be simply: from nothing!

INFLATIONARY COSMOLOGY: SOMETHING FOR NOTHING?

A modification to the standard 'big bang' view was proposed by Alan Guth in 1981 in his inflationary cosmological theory; during the 1980s this theory was refined by several cosmologists, including Andrei Linde and Paul Steinhardt.[13] Guth and the other cosmologists were aware of a number of physical problems which the standard view does not really resolve. Most of the important questions concern the initial conditions in the early universe. Why are the early conditions so smooth? Why does the expansion from the initial singularity proceed at the observed rate? What controlled the explosive force of the big bang? Why did the explosion and expansion happen anyway?[14] The inflationary theory tries to answer these questions in a natural and convincing way, and in doing so the theory makes the apparently outrageous suggestion that matter is created from 'next to nothing'.

The central idea of the theory involves a very rapid 'inflationary' expansion fuelled by massive repulsive forces in the early universe. Guth says that the inflationary universe develops from the initial singularity, but remains vague as to the earliest conditions. He admits that 'as with other cosmological scenarios, the starting point is a matter of taste and philosophical prejudice' – given the fact that the conditions of the rapid expansion in the inflationary universe prevent us from obtaining any substantial information about the pre-inflation universe.[15] However, he says that the inflationary theory starts from two simple and reasonable assumptions: the early universe should be extremely hot (over 10^{27} degrees Kelvin) and at least some regions of spacetime should be expanding rapidly and therefore cooling as they expand; he also estimates that material (in the form of energy) amounting to just 10 kilograms would be sufficient to provide the basic ingredients for the entire universe.[16] The figure of 10^{27} degrees Kelvin is important because it is supposed that at this temperature the universe in a normal vacuum state would undergo a phase transition. Above 10^{27} degrees Kelvin, fundamental nuclear and electromagnetic forces and particles are thought to be indistinguishable: particle physicists talk of a high degree of 'symmetry' between forces and particles and think in terms of 'grand unified theories' which are needed to characterise the physics involved in such 'uniform' situations.[17] However, without some mechanism to delay the phase transition, at this key temperature the symmetries are broken and the fundamental forces and particles become distinct. As the universe cools, these distinct forces and particles produce the more familiar material situations which we now observe around us.

Noting that material or a matter field in any given region may be characterised in relativity theory in terms of its energy density, Guth follows particle physicists in highlighting a very strange state for material possible in the early universe: the 'false vacuum'. We should not be misled by the use of the term 'vacuum': such regions are not invariably empty. Vacuum states describe the lowest possible energy density levels for a field in a given region of spacetime at a given temperature. Even in the lowest possible energy state, that of the true vacuum, there may be non-zero matter fields. Although the average value of a field in a particular region may indeed be zero, quantum theory predicts that there will be fluctuations around this zero value. Each fluctuation signifies the brief appearance of a 'virtual' particle. Hence, even in a true vacuum, matter fields may appear briefly. Even if the matter fields involved in the vacuum state are rather peculiar and certainly not observable in the sense that 'real' particles are, it is a mistake to think of any physical vacuum as some absolutely empty 'void'.

If the true vacuum state in the very early universe is maintained everywhere, then during the symmetry-breaking phase transition large numbers of exotic particles, known as magnetic monopoles, would have been produced. But there is no evidence to indicate that this did happen. So something must have prevented the phase transition from taking place in the true vacuum state. Inflation theory suggests that at least some of the energy in spacetime became fixed at a higher but unstable energy level: that of the false vacuum. In the false vacuum state, energy is able to postpone its transition from the highly symmetrical situation described in grand unified theories to a 'normal' situation which may be described in terms of more familiar electromagnetic and nuclear particles and forces. During the delay, the temperature of material in the false vacuum state 'supercools' to a temperature below that of the phase transition. This delay is critical in the overall development of the universe: during a short but eventful period, massive repulsive forces are produced which act on newly produced particles and antiparticles:

> While the inflationary model does not attempt to explain the formation of the initially hot expanding regions which subsequently supercool into the false vacuum state, it does explain the origin of most of the momentum of the cosmic expansion: the big bang gets its big push from the false vacuum.
> (Blau and Guth 1987: 543–4)[18]

The physical properties of vacuum fields in general and the false vacuum in particular provide the key to understanding inflationary theory:

1 In the false vacuum, the fundamental symmetries remain unbroken for at least some time, even below the phase transition temperature of 10^{27} degrees Kelvin. Although this unstable situation cannot be maintained indefinitely, advocates for inflation theory say that it holds just long enough to trigger both a massive expansion and the production of an enormous amount of material.
2 The energy density of the false vacuum is tremendous. 'It is the energy density that a large star would have if it were compressed to the size of the proton.'[19] This high density derives from the excess energy in rapidly cooling regions in which the symmetries between the fundamental nuclear and electromagnetic forces have yet to be broken. The situation here may be compared with the excess energy in supercooled liquids held at temperatures below their freezing-points but in which the general symmetries associated with liquids have not broken down to the less symmetrical state of a solid.[20]
3 The energy density of the false vacuum is fixed. For it is the minimum value for energy in that state. Hence, if the region occupied by a false vacuum expands, then the energy associated with that region must increase: more false vacuum, same energy density, therefore more energy overall contained within the false vacuum.
4 The pressure in the false vacuum is very large indeed and *negative*. A region with negative pressure has interesting gravita-

tional properties. The net effect of such a pressure is to produce a tremendous *repulsive* gravitational force. Hence an already expanding universe is given an 'inflationary' boost. During the false vacuum state, the universe doubles in size every 10^{-34} seconds.

5 The false vacuum state is unstable: as the temperature falls, the symmetries between forces and particles are more likely to be broken. Then the massive amount of energy stored in the false vacuum state is released. This energy release is typically in the form of particles and antiparticles. This material moves rapidly apart, given the enormous repulsive gravitational forces inherent in the false vacuum state. The observed global expansion of the universe is the result of this initial 'inflationary' push.

Perhaps the most celebrated aspect of the inflationary model is its assertion that the universe is a 'free lunch', that all matter was created literally from nothing: a creation *ex nihilo*.[21] But there is sometimes equivocation about whether the lunch is free or not, and Guth wisely adds the rider 'from next to nothing' when discussing the source of material in the universe. But how could the observed matter and energy content appear even from next to nothing? The above properties of the false vacuum give us the clue we need to solve this puzzle.

The energy available from the false vacuum state seems to be arriving from 'nowhere' and at no energy cost! What is happening may be described in a straightforward physical way, given the properties of the false vacuum. The total energy involved in both the production of matter and the fuelling of inflationary expansion is literally zero. Yes, the amount of energy needed to create the tremendous quantity of material in the universe is an enormous *positive* figure. But the amount of energy associated with the gravitational repulsion is an equally enormous *negative* figure. These two figures balance and cancel exactly. So the false vacuum provides 'equal and opposite' resources: positive energy for the creation of material in the form of particles and antiparticles; and an equal amount of negative (gravitational) energy which produces the rapid inflationary expansion. So we may say that the observed universe has its source in the false vacuum, which seems pretty close to next to nothing. The appearance of matter is in a sense a free lunch, for the false vacuum picks up the bill for matter and pays in credit from the friendly gravitational bank! But we should remember that Guth admits that we need about 10 kilograms worth of energy in the universe before the expansion and matter production associated with the false vacuum really gets under way – we might think of this as an energy cover charge. But Guth even suggests that the initial 10 kilograms of field energy may itself be balanced out by an equal and opposite amount of gravitational energy. There would then be no cover charge, and the total energy cost of the universe would be precisely nothing.

May we say that the inflationary universe is created from absolutely nothing, that inflationary theory gives us a model for creation *ex nihilo*? This would be unjustified. For, even if the total energy *cost* for the universe is zero, this does not imply that the universe 'starts' from an absolute void. After the initial singularity and before the inflationary expansion, there was a fluctuating and expanding vacuum field. Even if the average energy content of the field is zero, there is still a field with all its quantum fluctuations. A vacuum field may not be much, but it is something! The cause of the expansion and the production of material in the inflationary universe is clearly physical: there is no need to invoke some external non-physical 'divine' cause.[22] To ask for the physical cause of the initial vacuum field is inappropriate. All inflationary theory allows us to say is this: from the initial singularity until the false vacuum state, there was a slowly expanding and fluctuating (and therefore hot) vacuum field. The 'source' of this expansion lies at the singularity. But, as in the standard big bang model, if the singularity is a definite boundary to spacetime, there is no question of an earlier moment in time; and, if spacetime is open at the initial singularity with this singularity as a limit, then there may be no beginning to time at all. As we noted in the previous section, there is no *need* to seek some metaphysical explanation for the universe: we may regard it is as simply an existing self-contained entity. But, as before, there is nothing to prevent us from invoking some divine agency, except perhaps considerations of simplicity.

Many cosmologists regard inflationary theory as sound because it provides some very plausible answers to the many questions not tackled effectively by the standard big bang model. Some of these questions have already been tackled above. Others include:

1 Why is the universe so smooth globally? During the inflationary phase, as Stephen Hawking tells us: 'any irregularities in the

universe are smoothed out by the expansion, as the wrinkles in a balloon are smoothed away when you blow it up'.[23] The beauty of this idea is that it allows cosmologists to place far less emphasis upon the initial conditions holding before the phase transition: so long as we accept Guth's basic assumptions (that the universe is hot and expanding with unbroken symmetries between particles and forces), however irregular the early universe may be, the inflationary phase leads to the same generally smooth end result. Hence, there is no need to provide a detailed account of the very early universe. The formation of galaxies may then be explained in terms of local fluctuations in the globally smooth environment.

2 Why is the observed expansion rate so close to that for a 'flat' universe? This arises from the perfect balance between the energies involved in (i) the production of material and (ii) the gravitational repulsion. There is just enough energy in the form of gravitation to bring the expansion to a halt at 'infinity': i.e. the density of matter in the universe is not sufficient to bring about eventual recollapse, nor is it small enough to allow an accelerated expansion.

Variations on the theme of inflation also tackle questions such as the uniqueness of the universe and the necessity of a three-dimensional world. Andrei Linde's chaotic inflation theory suggests that the observed universe is just one domain or 'bubble'. Each domain is really a 'mini-universe' within a vast sea of bubbles which constitutes the universe proper. Inflation isolates the domains from each other, and our observations are restricted to our own domain. And only in certain domains will the orbits of electrons around atoms or planets around stars be stable: i.e. in domains with three spatial dimensions.[24]

BLACK HOLES

The thinking behind black holes is not a recent invention. The mathematician Laplace discussed, in 1799, the possibility of a star's gravitational attraction preventing light from leaving its neighbourhood, saying that 'the attractive force of a heavenly body could be so large, that light could not flow out of it'.[25] Laplace argues that this could happen if a star with a density close to that of the Earth were roughly 250 times the diameter of the Sun. And, in 1932, the Indian physicist Chandrasekhar tried (and failed) to persuade Sir Arthur Eddington that gravitational collapse to a 'singularity' could not be ruled out in a relativistic universe.[26] Furthermore, the idea of a universe expanding from an initial 'big bang' demonstrated that GTR may involve singularities: the initial conditions in the big bang scenario are singular because, as the volume of the universe decreases, the density of the (fixed) matter content approaches infinity. But only since the work of Stephen Hawking, Roger Penrose, Robert Geroch, and others in the 1960s have we had a detailed appreciation of the properties of black holes. Black holes are singularities of gravitational collapse. Wheeler coined the phrase 'black hole' to capture the idea that nothing could escape the gravitational attraction of the singularity. They are singularities because, as the surface area and therefore the volume of a collapsing star approaches zero, the density of the hole approaches infinity. An object with infinite density is a singularity and, as we have already noted in this chapter, a singularity is generally understood as the terminus for geodesics.

An important feature of the black hole is its event horizon. When a star is collapsing, there will be a critical stage in its collapse which represents the point of no return. This stage is reached when the star has a diameter and mass which are sufficient to induce total collapse by themselves. Before this critical stage is reached, there may be just a chance that the star may experience some internal or external stimulus which prevents total collapse. But, once the critical values for size and mass are reached, complete gravitational collapse is guaranteed. The diameter associated with this critical stage represents the diameter of the black hole's event horizon. When any particle enters the region defined by the event horizon, we may no longer see what happens to the particle. We have no way of detecting events within the horizon. For any such particle is then drawn inexorably towards the hole, and any signal (e.g. light) which might be used to transmit information about the progress of the particle is likewise drawn inwards. Hence, all information about the particle is cloaked by the horizon.

Stephen Hawking has shown that black holes do not simply grow fat on the material drawn into them by gravitational attraction. They eject particles in order to maintain equilibrium. The black hole acts like a strange kind of pump, sucking in nearby material by 'classical' gravitational attraction and ejecting it to 'infinity' via quantum processes:

> One can think of the emitted radiation as having come from inside the black hole and having quantum mechanically tunnelled through the potential barrier around the hole created by the gravitational field, a barrier that could not be surmounted classically; ... it is possible for a black hole to emit a television set or Charles Darwin ... but the overwhelming probability is for the emitted particles to have an almost thermal spectrum.
>
> (Hawking and Israel 1979: 19)[27]

The physics of this process is described in terms of three theories: GTR, quantum mechanics, and thermodynamics. Hawking suggests that the emission process sends particles and radiation to 'infinity', but Roger Penrose argues that a particle or group of particles might be intercepted just about anywhere since there is no way to rule out such particles having effects locally as well as at a distance.[28]

So the event horizon around a black hole acts as an effective barrier to any particle or signal trying to escape the gravitational pull of the singularity by classical means, but it does not block the progress of particles determined to find their way out through quantum tunnels. However, the event horizon has another role. Because physicists generally expect events close to the centre of the hole to be extremely hot, dense, and violent, they also expect that a quantum approach will be needed to explain what is happening inside the event horizon, just as they believe it will be required to give a full account of events close to the initial singularity. The horizon therefore is said to act as a classical cloak around quantum effects. Hence, classical (i.e. non-quantum) GTR cannot be used to characterise the singular behaviour inside the horizon. Consequently, GTR predicts phenomena which it cannot handle: in a sense, the theory predicts its own breakdown. But the existence of the horizon allows us to use classical theories like GTR up to the horizon and leave the problem of what is happening inside in abeyance.

COSMIC CENSORSHIP

Roger Penrose has suggested that the existence of horizons is not a happy accident, but a general consequence of the cosmic censorship hypothesis, which requires that external observers should always be protected from the quantum gravitational effects associated with black holes by event horizons. This hypothesis depends on the definition of a Cauchy hypersurface. So we shall explore the ideas behind this hypersurface before turning to an analysis of cosmic censorship.

In Chapter 8, we investigated various levels of causal structure which might be imposed on a spacetime. The strongest condition discussed there was that of stable causality. A causally stable spacetime is one in which closed timelike loops are impossible. But we may place further constraints on spacetime. In particular, we may demand that nothing should prevent us from predicting the future of a region of space in which no point in the space causally precedes any other point. Such a region is effectively a local 'time-slice', called an achronal slice. If we know all there is to know about the physical conditions on this slice and these conditions are not singular, then we may use the field equations to determine the causal future of this slice. Just as we may determine the future on the basis of the conditions on the slice and the laws which apply, symmetry considerations show that we may also determine the past. Clearly, as we move further into the future, the scope of our predictions grows ever smaller. This is because as we move along a world line passing through the slice, the chances of meeting a world line which does not pass through the slice increase. However, the possibility of meeting a world line from outside the slice is restricted by the fact that no causal signal may travel faster than light. So we may still construct a region in which all events may be determined: such a region is called the domain of dependence of the achronal slice; see Figure 36 (p. 213). If we can construct an achronal slice through the entire spacetime, then prediction is not restricted to just local events, but may apply to all spacetime. Such a global time-slice is called a Cauchy hypersurface and its domain of dependence is the entire spacetime. Therefore this global hypersurface seems to allow us to determine every event in a spacetime. This kind of determinism is linked with the idea of 'Laplacian determinism', a perspective suggested by Laplace in which a complete set of information about the present together with the true laws of nature could provide us a complete history (past, present, and future) for the universe.[29] Spacetimes which possess Cauchy hypersurfaces include Minkowski spacetime and the Friedmann models.[30]

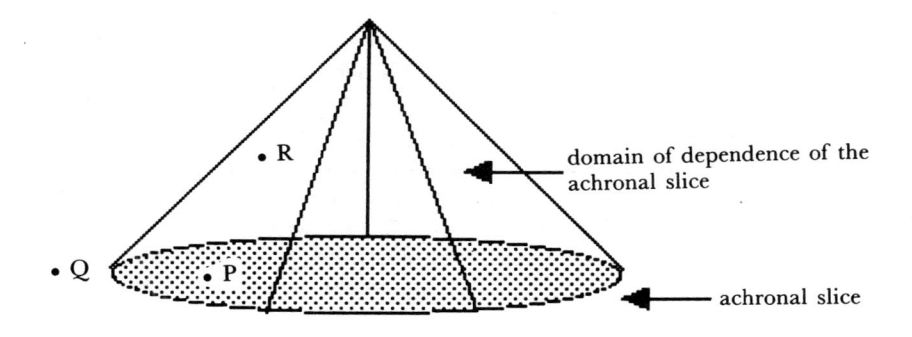

Although a world line can reach R from many points on the achronal slice, including point P, no signal from any point outside the slice, such as Q, can reach R – unless of course it travels faster than light! If we exclude this possibility then all events within the domain of dependence of the slice may be influenced only by events on the slice.

Figure 36 Achronal slice and domain of causal dependence

However, the Cauchy hypersurface and the deterministic equations of GTR may not be sufficient to guarantee complete predictivity in a spacetime by themselves. We noted in the final section of Chapter 7 that those who take a substantivalist view of spacetime points may be unable to hold on to determinism, given the 'hole argument'. But we must also face the possibility that any singularity without an event horizon may introduce indeterminism via quantum effects into the domain of dependence. Although we might stipulate that the conditions on an achronal slice be non-singular, a singularity of collapse may well evolve to the future of the slice, given the conditions on the slice. This is where the event horizon comes in. For an event horizon is supposed to shield spacetime from the quantum effects, so that the complete causal determination of events will not be disrupted. Therefore, Penrose suggests that the cosmic censorship hypothesis must hold in GTR if we are to have a chance of achieving complete predictivity.

Could we have a singularity without an event horizon? And would there be any sign of such a singularity in the physical conditions which obtain on an achronal slice? It turns out that there are two main ways in which this may be possible:

1 Chris Clarke and others have shown that naked singularities may evolve in an apparently well-behaved spacetime; these are physical singularities without event horizons. Particles may emerge in a non-predictable way from naked singularities to our future and interfere with events, thus disrupting attempts to provide a complete causal determination of the future; symmetry considerations show that we would likewise be hindered in any such attempt to determine the past fully. Clarke has shown that naked singularities may develop to the future of a non-singular achronal slice.[31]

2 There may be topological holes present in a spacetime – a spacetime with regions literally cut away (do not confuse them with the 'holes' in the hole argument of Chapter 7); although these are not physical entities in the sense of naked singularities, they may be the starting-points for non-predictable geodesics and so they could be just as causally disruptive as naked singularities.[32]

We might decide to rule out such entities on empirical grounds: there seems to be little or no empirical evidence for the existence of naked singularities – all they have is theoretical backing; and topological holes might be dismissed as being no more than mathematical curiosities with the weakest of empirical foundations. Such decisions would be in line with the cosmic censorship hypothesis. Therefore, the claim that cosmic censorship holds is essentially empirical. It is motivated by the desire to maximise causal determination within the context of GTR.

However, we cannot rule out the initial singularity and black holes on empirical grounds so easily. And the big bang and black holes may produce effects which threaten causal determination almost as much as naked singularities and topological holes. If quantum effects dominate the times just after the big bang, then GTR's field equations cannot tell us what is happening then. If black holes may eject material into spacetime quantum-mechanically, then GTR's field equations cannot help us predict with certainty what lies to the causal future of a black hole. To ensure complete causal determination in spacetime, we would have to rule out *all* singularities and advocate a 'strong' version of the hypothesis ruling out anything likely to interfere with causal determination; and, if we take the hole argument in Chapter 7 seriously, absolutists would also need to revise their attitude towards spacetime points. Given that we have good empirical grounds for the initial singularity and for black holes, there seems to be little reason to push cosmic censorship to such limits. But, if we apply censorship to just naked singularities and topological holes, then the hypothesis in this 'weak' form applies to just those things we find empirically objectionable; and, in doing so, the weaker hypothesis fails to achieve the goal of the cosmic censorship: causal determination. The hypothesis might be a useful way of grouping those solutions of GTR which are 'classical' in every way; and it may help us to explore the general properties of such solutions. But there seems little reason for us to insist that GTR should be limited by the hypothesis in either its weaker or stronger form. The field equations of GTR may be deterministic in a classical sense, but the phenomena which they predict and with which they deal are sometimes indeterministic. We should not drive a wedge between the theory and its theoretical and empirical context, no matter how nicely the cosmic censor asks.

DETERMINISM VERSUS INDETERMINISM

Singularities and the causal disruption associated with singularities may turn out to be explained fully in terms of the material properties of objects in spacetime. They do not pose any immediate threat to a relationist position. But a topological hole represents an irreducible element of spacetime structure; and such holes in spacetime may be just as causally disruptive as physical singularities. To accept the possibility of topological holes is to concede that spacetime may have structure over and above its material contents. Such holes are likely to have direct effects on the affine structure of spacetime, since they are the starting- and finishing-'points' for objects moving along the geodesics of spacetime; and, as we saw in Chapter 7, how objects move in spacetime is defined in terms of the affine structure. Therefore the recognition that a successful theory of gravitation must deal directly with indeterministic phenomena may lead to the concession that the spacetime of such a theory is likely to be essentially absolute in the sense that its structures may not be completely reducible to its material contents. So a relationist would be eager to rule out topological holes as candidates for existence.

However, relationists do not seem to have a particularly strong argument for this move: if they say that we must deny their existence because such holes are yet to be observed, then they must also ask us to rule out a tremendous variety of theoretically respectable entities as candidates for existence; if they adopt a determinist position and urge some version of the cosmic censorship hypothesis, then we would be asked to neglect a number of plausible gravitational phenomena simply because they are tarred with the same indeterministic brush as topological holes. The problem here as elsewhere for the relationist is that the structures of spacetime are rich enough and complex enough to allow us to generate a range of interesting features which are characteristics of spacetime itself. Absolutists have the seemingly limitless ingenuity of the mathematician on their side. Relationists must resort to often ineffectual complaints about the liberties taken by mathematicians when dealing with the physical world.[33]

This leads us to an interesting problem concerning the hole story. John Earman says that taking spacetime points too seriously leads the absolutist to a dilemma. We have noted that those who are substantivalists about spacetime points must concede the possibility of a single Cauchy hypersurface allowing the construction of alternative model spacetimes *with different futures*. But this means that determinism is, at least for substantivalists, a lost cause – for the information specified on the hypersurface does not allow us to predict the future unambiguously! But, if we take spacetime structure seriously, then we are also likely to have no fundamental objections to topological holes and other indeterministic phenomena. So an absolutist perspective on spacetime structure seems to lead directly towards indeterminism. Earman tries to persuade those with absolutist inclinations to drop their substantivalism about points and to follow the path suggested by Sklar's notion of 'absolute motion without absolute spacetime'. This might seem attractive to those who dream Laplacian dreams. However, those of us who are convinced that the world is essentially indeterministic are more likely to continue to use those models and structures which enable us to describe this apparently indeterministic world as coherently as we can. We do so, not because of any absolutist prejudices, but because the empirical evidence suggests that the world is essentially indeterministic.

CONCLUSION: RELATIVITY – JUST ANOTHER BRICK IN THE WALL?

INTRODUCTION

One major problem for the philosopher of space and time is the fact that scientists keep changing their minds. They do so when they make 'revolutionary' leaps from theory to theory; and they continue to do so as they develop the ideas of a given theory. Sometimes the end product of such a development might be just as revolutionary when compared with the original formulation of the theory as an entirely new theory. Sometimes the changes made are not quite so dramatic. All this leads us into difficulties. Which set of ideas should we investigate? Which theory tells us about the space and time of the actual world? Should we accept that we are part of a process leading towards the truth, but that we have a long way to go? Should we accept Stephen Hawking's suggestion that the end of theoretical physics may be in sight?[1] Or should we take each theory with a pinch of salt: just one more invention of some rather brilliant mathematicians with nothing to say about the real world? Any attempt to answer such questions takes us firmly into the territory of the philosophy of science. But our deliberations about the nature of space and time may help us to understand how we might approach such questions.

Nobody is likely to claim that relativity theory tells us the whole truth about space, time, and motion. There are too many unsettled scientific problems, too many anomalies, for anyone to be so bold (or so foolish). However, when we begin to investigate the structure and contents of relativity theory, we start to find that relativity is not some monolithic, moribund dogma. It is a vibrant, changing, wide-ranging context in which empirical, theoretical, and philosophical issues are subject to continual debate. But this should make us all the more cautious in trying to give final answers to many of the problems of space and time. We may certainly claim 'relativity interpreted in such-and-such a way has such-and-such implications for the nature of space and time'. We may not conclude that any such assertion is the final word about space and time. Of course, we may regard the empirical success of relativity as evidence for at least its approximate truth; we may think that the theory captures something of the physical world. But relativity theory is not alone in this: other theories involving space, time, and motion seem to apply to the physical world in much the same way, if not always with the same degree of success.

Is relativity just one more theory, just one more brick in the wall of science? Will its fate be banishment to the archives? Or should we see it as part of a long-term evolution, with its essential ideas involved not just in the Special and General Theories of Relativity (STR and GTR) as originally conceived, but also in the long history of thought on space, time, and motion: past, present, and future? How we answer such questions depends upon our view of spacetime theories in particular and scientific theories in general. And, with a clearer view of the structure of theories, we may become even more cautious before providing final answers to the many questions concerning space and time before us: the status of the continuum; the problems of conventionalism; the rationale for absolutism and relationism; the issue of time travel; the beginning of time; and the nature of singularities.

WHAT IS A THEORY?

Physical theories are frequently characterised as structures built around stable fundamental laws. Whether we adopt Kuhnian, realist or falsificationist views of science, the laws of a theory are seen as the focal points of the theory and consequently philosophical investigations and reconstructions have typically concentrated on the central role of fundamental laws in physical theories and the implications of those laws. However, many writers now warn against placing undue emphasis upon the theoretical context and the theoretical laws of science. Cartwright, Galison, and Hacking attack the primacy of theory and direct us to the experimental context in which the often clumsy concoctions of the experimentalists replace the elegant generalisations of the theorists.[2] Philosophers have indeed neglected the experimental domain and

CONCLUSION

it is right that this should be rectified; but we should be careful not to under-emphasise the interactions between theory and experiment. If we perceive a theory as consisting in just a law or set of laws, then it is easy to drive a wedge between theory and experiment. However, we may adopt a view of physical theories which will help us to provide a bridge between those who see theory as the primary driving force of science and those who wish to emphasise the autonomous or at least partially autonomous and pivotal role of experimentation. The view which I shall discuss here draws on the ideas involved both in Thomas Kuhn's view of a disciplinary matrix and in Imre Lakatos' notion of a research programme.[3] But we shall find that there are a number of sound reasons why we should adopt an even broader perspective of physical theories.

We should avoid the temptation to characterise a physical theory as having a 'limited nature and scope' – instead we should regard a typical theory as a complex and wide-ranging theoretical structure.[4] If we fail to do this, then we may develop a rather limited philosophical appreciation not only of the relationship between theory and experiment but also of such problems as the theory-dependence of observation and the continuity of theoretical knowledge. Mature physical theories are certainly not simple structures: they involve more than fundamental laws – this much has been learnt from such writers as Kuhn, Hesse, Lakatos, and Laudan, who variously direct us to the important roles of standard derivations and problems, models, methodological conventions, and values.[5] In particular, Kuhn's notion of an established theory (the disciplinary matrix) involves:

1 stable law-like symbolic generalisations which may also function as definitions of the symbols used in them;
2 epistemological (and possibly metaphysical) commitments to certain phenomena and entities and also to analogies, metaphors, and models;
3 methodological and pragmatic values; and
4 exemplars – the concrete solutions of central problems.[6]

And Lakatos makes an interesting distinction between the negative heuristic and positive heuristic of a research programme:

1 the negative heuristic is the unfalsifiable central 'hard core' of generalisations in a theoretical research programme so that the

development of a fruitful programme is strongly focused by a firm commitment to the laws of the theory; to reject this hard core is to reject the entire theory; hence it is the negative heuristic which underwrites the stability and continuity of a given programme;

2 the positive heuristic acts as a 'protective belt' of supporting assumptions and ideas around the hard core; when a theory is challenged, the auxiliary assumptions in the positive heuristic are typically modified to help maintain theoretical stability.

The theoretical contexts of GTR, Quantum Mechanics (QM), and other established theories exhibit, *to a certain extent*, these Kuhnian and Lakatosian features. In fact, a thorough review of any established theory will reveal a tremendous array of possible constituents.

Peter Galison says that the stability of science derives from the amorphous nature of the enterprise.[7] A crystal is easily split apart; but amorphous materials like fibreglass, with its distinct but interweaving threads, are much more robust. He says that science is like a brick wall: with three kinds of interlocking layers – theoretical, observational, and experimental. Each of these three domains of the scientific enterprise have partial autonomy from the other two. This appears to be correct. We may certainly identify peculiarly experimental features or theoretical aspects of any major physical discipline like gravitation. But we must also recognise that there is a high degree of interaction between these three domains. Hence, it may be a mistake to focus as strongly as Kuhn and Lakatos sometimes seem to do on the central theoretical generalisations of a theory as the load-bearing structures in the scientific enterprise. Observation and experimentation also play important parts in providing stability and continuity.

Galison's general view of science may also apply to each individual domain. Each domain may be regarded as an amorphous structure, with a vast array of interlocking elements and ideas. We have seen that a theory like GTR is a composite structure. And its internal stability is provided by the interaction of all its various elements through time. No element is immune to change, not even the central laws. Those elements with the greatest stability tend to be: general principles of symmetry and invariance, which may apply equally well to alternative theories; and those ideas which are common to several dominant theories of the day.

CONCLUSION

So Galison's metaphor of a wall is illuminating but perhaps too simplistic. Instead we might regard GTR as a part of a strong rope of inter-weaving threads: some threads are short, some are long. The rope is science. The threads are the constituent elements of the interacting domains of observation, experiment, and theory. The strength of the rope lies in the weave of all these diverse elements.

We need to investigate the ideas involved in a theoretical context. This will enable us to capture something of the nature and scope of physical theories, especially those concerned with space and time, which provide our main focus here. This should help us to see a little more clearly why science is an essentially stable enterprise. But it will also help us to see why we should be cautious in our attitude towards the problems of space and time.

THE STRUCTURE AND SCOPE OF SPACETIME THEORIES

There are six main ingredients of physical theories: laws; principles; epistemological commitments; analogies; the resource base; and values. We shall consider each in turn and see that each element plays an important part in spacetime theories such as Newtonian gravitation and GTR.

1 Laws. There are three main types of physical law:
   a Fundamental laws such as the Einstein field equations of 1915-16.
   b Laws derived from the fundamental laws using specific initial and boundary conditions to focus on a particular problem: such as the equations used in the Schwarzschild solution of GTR to characterise the gravitational field around a single, isolated mass. Such derivations are frequently used to generate observational and experimental predictions.
   c Inter-theoretical laws which draw on more than one theory to help us characterise either a theoretical or an experimental context in as realistic a manner as possible. The Hawking–Penrose theorems of 1970 and the Hawking equations of 1974 help us to theorise about the behaviour of black holes by drawing on thermodynamics and QM as well as on GTR. And the equations produced to help us estimate the effects of gravitational waves on large aluminium cylinders – typical features in the early experimental apparatus used in the search

for such waves – take into account the bulk elastic properties of materials.

2 Principles. There are three main kinds of principle – but the divisions between each kind are not so clear-cut as in the case of laws:

  a Empirical principles. For example, the principle of conservation of local energy-momentum is an essential part of GTR just as the conservation principles for energy and of momentum are deeply embedded in Newtonian theory.

  b Methodological principles. Principles of equivalence, invariance, and general covariance acted as a guide to Einstein in his search for the field equations of GTR. And symmetry principles play an important methodological role, for example, in helping us to capitalise on symmetrical features of models of GTR such as Schwarzschild's (spherically symmetric) solution.

  c Philosophically motivated principles. The Strong Anthropic Principle and Mach's Principle are usually regarded as philosophical principles, based perhaps on some vision of the role of consciousness or a particular philosophical perspective, like positivism.

It is hard to make a clear distinction between these three kinds of principle, for methodological principles frequently have strong empirical pedigrees, as is the case with the principle of equivalence, and they occasionally have philosophical overtones, as is the case with the principle of general covariance, which demands 'simplicity' in the formulation of GTR's equations. Again, there are hard choices in cases such as the cosmological principle, when it is difficult to decide whether its high empirical standing completely outweighs its Copernican philosophical 'prejudice' against any form of anthropocentricity.

Indeed, the situation is complicated further by the fact that the status of principles may change during the development of a theory – as indeed happened in the cases of both Mach's Principle and the Cosmological Principle, the former gradually falling out of favour and the latter rising in status.

3 Epistemological commitments. We just cannot make sense of either our laws or our principles if we fail to give some meaning to the elements which they contain. GTR contains a range of basic 'building-blocks' which enable us to interpret the various laws of the theory; these are:

CONCLUSION

a Fundamental commitments to entities. Commitments to basic geometrical conceptual 'building-blocks' are needed to construct the metrics and energy fields of GTR spacetimes. When Locke compiled his list of primary qualities, he provided a review of the basic commitments which he saw as essential to the mechanical perspective inherent in Newtonian theory. Similarly, we need to assimilate such basic geometrical elements as the affine connection if we are to operate within GTR's context.

b Commitments to constants used in the laws of the theory. Sometimes these commitments are deep-seated, as in the case of the Newtonian gravitational constant or Planck's constant. Occasionally there are mixed reactions to a constant: as in the case of the cosmological constant, which Einstein first added to his field equations in 1917 and later dropped when empirical evidence called its use into question.

c Commitments to phenomena or entities which appear in a wider scientific context. GTR's commitment to a finite speed *in vacuo* for light which is constant for all inertial observers is also a basic demand in both STR and electromagnetism. Such commitments need not always be at a high theoretical level: we may also hold lower-level 'observational' commitments, for example, to the 'fact' that locally light travels in straight lines.

Of course, these are not the only epistemological commitments made in a theory. We will also be committed to laws, principles, models, and so on. But what distinguishes the fundamental commitments listed above is the fact that they are epistemologically prior to the laws, principles, and models of the theory. Without such basic commitments, we could neither use nor grasp (let alone develop) a theory.

4 Analogies. There are three basic typical kinds of analogy to be found within theoretical contexts:

a Basic metaphors are particularly useful in theories for several reasons. First, they provide links between our macroscopic appreciation of the world and our microscopic perspective, allowing us to grasp and elucidate the latter by reference to the former; for example, when we use the idea of 'spin' to characterise the behaviour of particles in QM, we may begin to explain the idea by pointing out that two identical rotations through 360 degrees for a person would be two distinct rotations for a particle with spin 1/2. Secondly, they can help to stimulate imaginative research within a theoretical context; for example, when similarities are spotted amongst situations in distinct theoretical contexts, as is the case in Hawking's analysis of black holes, which he treats as a kind of thermodynamic 'pump' absorbing particles classically and then emitting them quantum-mechanically. Thirdly, a context may be fruitfully dominated by a single metaphor, as in the ten-dimensional superstring theory of Green and Schwartz, in which different particles are depicted as different modes of vibration on a fundamental 'string'.[8]

b Physical models can help us to understand unfamiliar concepts. The 'expanding balloon' model of modern cosmology offers a way of visualising an expanding spatially closed universe, as in the case of basic metaphors. Such models frequently provide a link between problems within the theory and more familiar physical contexts.

c Conceptual models frequently have a central role in helping to develop a theoretical idea. Such models allow us to explore the implications of those high-level theoretical ideas which are remote from the domains of observation and experiment: for example, the inflationary cosmological model of the early universe, in which an initial rapid expansion is driven by the energy inherent in the vacuum state – an expansion from as near to nothing as we can get.[9]

Models and metaphors may have their limitations in so far as: first, they focus on the key positive analogies and often push to one side negative aspects which might nevertheless be of critical importance; and, secondly, they introduce a markedly imaginative, intuitive, and therefore, in some eyes at least, undesirable subjective element into theoretical contexts. However, without models and metaphors, theories and theorising would be severely impoverished.

5 Resource base. In order to operate within the theoretical context effectively, more is needed than models, metaphors, and basic epistemological commitments. We need a resource base which contains the essential tools required to develop and to enrich the theory. Such a basis allows us to explore the range of application of the theory and to project the theory into both familiar and unfamiliar physical situations. The resource base of physical theories typically includes: abstractions; solved problems and standard applications; and thought experiments:

CONCLUSION

a Abstractions allow scientists to focus on problems, helping to avoid too many distractions. When we use the key idea of a local inertial frame in GTR, we consider the 'limiting' situation of an infinitesimal region. Without such simplifying manoeuvres, it is hard to see how we could plough through what would be a mathematical quagmire: GTR would be practically useless. Furthermore, some abstract mathematical models take on a central role in the resource base, for example, Friedmann's cosmological solutions of Einstein's field equations, which depend upon the assumption that the universe is everywhere filled with dust, are regarded as the yardstick for any global view of the universe.

b Solved problems – Kuhn's exemplars: GTR includes a number of successful concrete solutions to pre-existing problems such as the account of Mercury's orbit and also to new problems such as the phenomenon of red shift.

c Mathematical tools are of course essential for any physical theory. It is interesting to note that a theory such as GTR may be formulated mathematically in several ways. For example, we may produce Lagrangian or Hamiltonian formulations of GTR, which are particularly useful in addressing quantum gravitational problems.[10] Hence, a theoretical context may be enriched by having access to a range of mathematical tools. Indeed without such classical mathematical tools as canonical formalism, the development of QM would have been seriously impeded.[11] Elie Zahar also reminds us of the importance of formalism when he argues that many interesting contributions to relativity (STR and GTR) were stimulated by the conceptual potency and potential of the available mathematical techniques, which acted as a powerful heuristic.[12]

d Thought experiments often play a pivotal role in spelling out theoretical ideas. Sometimes thought experiments help us to visualise ways in which a theory may be tested. Sometimes they help us to make a conceptual point about a theory, as in the case of Newton's rotating bucket experiment. And sometimes they may be used to indicate contradictions or incoherence within a theory.[13]

6 Values. These ingredients of a theoretical context are supplemented, informed, and constrained by commitments to a range of internal and external values: methodological values of objectivity or of verification and falsifiability or of simplicity;

pragmatic values of economy or of usefulness; motivational values which may drive individual scientists in their search for truth or coherence or consistency or theoretical unity; social, ethical, and personal values which may bind and direct the scientific community; and sometimes 'metaphysical' values which may influence a scientific discipline for a time, such as the conviction behind the Anthropic Principle that humanity has a fundamental place in the cosmological 'scheme'.[14]

THE LAST WORD?

Both Lakatos and Kuhn regard fundamental laws with a good deal of reverence. They are seen as focal points of a theoretical context. But we have strong reasons for taking laws a little less seriously:

1 The complexity and diversity of a theoretical context such as GTR indicate that fundamental laws should be regarded as an integral part of such a context.
2 We must recognise that even fundamental laws have explicit restrictions placed upon them: the field equations of GTR are subject to energy and causal conditions which focus them upon physical possibilities rather than mathematical curiosities.
3 And fundamental laws also seem to have implicit ranges of application: there is no expectation that the laws should apply in all physically possible situations – no one expects GTR to apply accurately to the very early universe or to conditions within a black hole.
4 Principles of invariance and symmetry seem to dominate the theoretical contexts of physics rather more than fundamental laws as such: ideas of symmetry, covariance, and invariance allow us to focus upon the conservation of quantities in an extremely general way.[15]
5 Fundamental laws are not static museum pieces: they may change in form, in content, and in their range of application over time. The laws of Newtonian mechanics, as we noted in Chapter 9, have gone through a number of metamorphoses; and the field equations of GTR have not been as stable as some would like to think.
6 As Nancy Cartwright points out, fundamental laws are not strictly speaking true of anything in the world: they may be accurate within their abstract domains, but phenomenological

CONCLUSION

and experimental 'laws' are typically used to capture fully the character of the physical world.[16] Hence, fundamental laws should be seen as partial ingredients for our, frequently intertheoretical, phenomenological descriptions of the world.

7 Too much emphasis upon fundamental laws isolates physics from other scientific fields. Many theories in other fields, such as geology or biology, share many of the features of the theoretical mix described above. But they rarely have anything like fundamental laws. If Cartwright is right in thinking that fundamental laws cannot be true, then it may be foolish to dream of the day when we can provide fundamental laws for every field of science.

These considerations suggest that we should adopt a sceptical attitude towards any claim for the pre-eminence of the laws of a theory. Laws do have a role in science; but this role is within the context of science as a whole. This context is an evolving, historical process. Hence, any conclusions we draw about theories must take this into account.

GTR, like other physical theories, has developed and changed its character many times and in many ways since its original development. The form and content of the field equations have changed and changed again. Constants have appeared and disappeared and reappeared. Classical approaches have given way to composite classical–quantum methods. Mach's Principle, the Anthropic Principle, and the Cosmological Principle go in and out of fashion. Such dynamism makes it difficult to resolve many of the questions raised in this book about the character of space, time, and motion. We might point to a particular isolated part of a time-slice of the theory and say: yes, Einstein's theory (as interpreted by physicist X in the 1970s) commits us to spacetime as an irreducible element in our description of gravitational phenomena; or yes, the cosmic censorship hypothesis applies to GTR conceived classically. But, as soon as we see a theory of space and time as an evolving theory set in a more general evolving scientific context, if such assertions are meant to apply to the entire dynamic context, then they are rather less than accurate; see Figure 37 (p. 228).

Too much attention to any one narrow characterisation of a theory such as GTR carries with it not just the danger of inaccuracy. We may also be led into a dogmatic frame of mind. Without diversity within a theoretical context, without

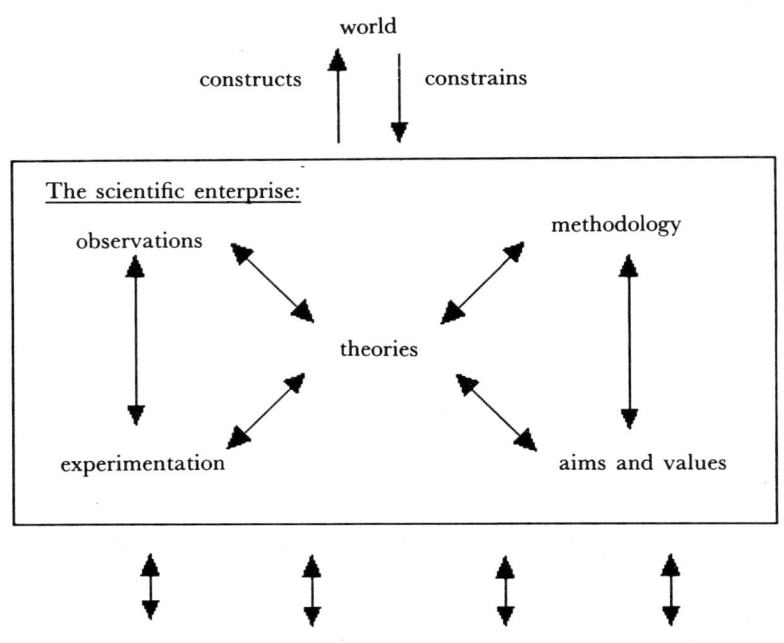

```
                        world
          constructs  ↑ │ constrains
                      │ ↓
┌─────────────────────────────────────────────────┐
│ The scientific enterprise:                       │
│    observations              methodology         │
│           ↖   ↘      ↗   ↙          ↑           │
│           ↕        theories         ↕           │
│           ↙   ↗      ↖   ↘                      │
│    experimentation           aims and values     │
└─────────────────────────────────────────────────┘
        ↕           ↕           ↕           ↕
```

General cultural beliefs, political and economic constraints, scientific institutions, technological base for science, educational practice, general values – all such influences constrain and inform the scientific enterprise

Figure 37 Theories in context

commitments to a variety of principles and procedures and formalisms, we run the risk of intellectual stagnation. Relativists are to be congratulated for the complex nature of GTR. Yes, individual scientists may have dogmatic commitments to a given point of view. But the entire community should be encouraged to remain as diverse as possible. GTR is a testament to intellectual diversity. Although we may be frustrated by the inability of the theory to give us clear-cut answers to many 'basic' questions, there is sufficient cohesion in the theoretical context as a whole to allow a reasonably straightforward account of many problems. That cohesion is typically not provided by any one element in the context. Our ability to address physical and philosophical issues in general depends on the continuity and stability which GTR provides as a dynamic theory in a wider scientific context.

NOTES

INTRODUCTION

1 The full text of this report appears in Pais, A. (1982) *Subtle is the Lord: the Science and Life of Albert Einstein* Oxford: Oxford University Press p. 307.

1 ZENO AND THE LIMITS OF SPACE AND TIME

1 Aristotle *Physics* 239 b 5 – 240 a 18 and 233 a 21–31.
2 See Salmon, W.C. (ed.) (1970) *Zeno's Paradoxes* Indianapolis: Bobbs-Merrill p. 34; and Penrose, R. (1971) 'Angular momentum: an approach to combinatorial space-time' in Bastin, E. (ed.) *Quantum Theory and Beyond* Cambridge: Cambridge University Press; and also Penrose, R. (1987) 'Newton, quantum theory, and reality' in Hawking, S.W. and Israel, W. (eds) *300 Years of Gravitation* Cambridge: Cambridge University Press, in which Penrose renews his interest in the non-continuous 'twistor' approach to spacetime.
3 See, for example, Peirce, C.S. (1935) *Collected Papers* Cambridge, MA: Harvard University Press; Peirce says that the Achilles paradox 'presents no difficulty to a mind adequately trained in mathematics and logic', vol. VI, p. 177.
4 Stewart, I. (1987) *The Problems of Mathematics* Oxford: Oxford University Press.
5 Owen, G.E.L. (1957) 'Zeno and the mathematicians' *Proceedings of the Aristotelian Society* **58** p. 199.
6 Salmon, W.C. (ed.) (1970) op. cit., p. 141.
7 Simplicius was a sixth-century AD commentator on Aristotle.
8 Barnes, J. (1979) *The Presocratic Philosophers* London: Routledge presents a detailed discussion of this paradox; see vol. I, ch. 12.
9 Cauchy, A. (1821) *Cours d'Analyse* Paris.
10 f course, if Achilles is running at a steady speed, his speed is always greater than that of the tortoise; and we can therefore predict that, when Achilles approaches the tortoise, he will indeed overtake it.
11 Thomson, J.F. (1954) 'Tasks and super-tasks' *Analysis* **15** pp. 1–13.
12 Sainsbury, R.M. (1988) *Paradoxes* Cambridge: Cambridge University Press.

13 Benacerraf, P. (1962) 'Tasks, super-tasks, and the modern eleactics' *Journal of Philosophy* **59** pp. 765–84; see also Berresford, G.C. (1981) 'A note on Thomson's lamp "paradox"' *Analysis* **41** pp. 1–7.
14 Sainsbury, R.M. (1988) op. cit., p. 1.
15 Black, M. (1950) 'Achilles and the tortoise' *Analysis* **11** pp. 91–101.
16 Salmon, W.C. (ed.) (1970) op. cit., p. 34.
17 Hesse, M.B. (1966) *Models and Analogies in Science* Notre Dame, IN: Notre Dame University Press provides a classic statement of the role of models in physical science.
18 Newton-Smith, W.H. (1980) *The Structure of Time* London: Routledge, pp. 68–73, 121–6.
19 The thesis may be traced to Pierre Duhem's ideas in his (1906) *La theorie physique: son objet et sa structure* Paris; the quotation here is from Quine, W.V.O. (1970) 'On the reasons for indeterminacy of translation' *Journal of Philosophy* **LXVII** 6 pp.178–83. See also Hesse, M.B. (1974) *The Structure of Scientific Inference* London: Macmillan, and Hookway, C. (1988) *Quine: Language, Experience and Reality* Cambridge : Polity Press.

2 CLOCKS, GEOMETRY AND RELATIVITY

1 The idea of an inertial frame of reference will be explained in Figure 4 (pp. 27–32) and in the following section.
2 Einstein's (1905) paper introducing STR appears in translation in Stachel, J.J. (ed.) (1989) *The Collected Papers of Albert Einstein vol. II The Swiss Years: Writings, 1900–1909* Princeton: Princeton University Press, English translation text by Anna Beck; see also Einstein, A., Lorentz, H.A., Weyl, H., and Minkowski, H. (1923) *The Principle of Relativity* London: Methuen; the original paper appears under the title 'Zur Elektrodynamik bewegter Korper' *Annalen der Physik* **17** (1905) pp. 891–921, also reprinted in Stachel, J.J. (ed.) (1989) op. cit. (main text), pp. 276–310.
3 See Bergson, H. (1976) 'Discussion of the paradox of the twins' in Capek, M. (ed.) (1976) *The Concepts of Space and Time* Dordrecht: Reidel.
4 We measure the lifetime of such particles in terms of their half-life – although we cannot say when any given particle will decay from its unstable form, we can say with high accuracy that a large group of such particles will be reduced to half the original number in a specific time and will be reduced to a very small quantity indeed within a definite period. For details of this experiment see Frisch, D.H. and Smith, J.H. (1963) 'Measurement of the relativistic time-dilation using mu-mesons' *American Journal of Physics* **31** p. 342; and also French, A.P. (1968) *Special Relativity* London: Van Nostrand Reinhold, which reports the experiment: pp. 98–9.
5 See Salmon, W.C. (1980) *Space, Time, and Motion* second edition Minneapolis: University of Minnesota Press for a more detailed report of this experiment.
6 Of course, we will only believe the predictions of the LT so long as the laws of physics are not violated in any way. STR needs to be consistent

with, for example, the accepted laws of electrodynamics. This constraint upon STR – the so-called Principle of Relativity, demanding that the laws of physics should hold good in all inertial frames – was one of the building blocks – of the theory.

7 One of the reasons for adopting a relativistic perspective is the fact that distant events cannot be said to be simultaneous given the lack of any signal capable of travelling between clocks instantaneously. The idea of simultaneity may only be attached to nearby events. So, because the time the light takes to travel from Big Ben to me is finite, when I synchronise my clock with Big Ben, 'my time' is always 'behind' by a factor related to the distance between me and Big Ben when the synchronisation took place. But time-lag is fixed; so long as I stay put my clock will be just as much behind Big Ben at the end of the hour as it was at the beginning.

8 Where the factor $\beta = (1 - v^2/c^2)^{-1/2}$. Because v cannot be greater than c, it is clear that β must be greater than or equal to 1.

9 So far I am allowing STR this luxury – but in Chapters 3 and 4, which focus on conventionalism, we shall examine the thinking behind the assumption that clocks in the same frame may be said to synchronise perfectly.

10 The spacetime interval (Δs) between two events is given by:

$$(\Delta s)^2 = (\Delta t)^2 - (\Delta x)^2 - (\Delta y)^2 - (\Delta z)^2$$

Δs therefore depends on the spatial distances ($\Delta x, \Delta y, \Delta z$ – the differences in the x,y,z coordinates) and the temporal duration (Δt) as observed from any inertial frame of reference (with units chosen so that c = 1). A finite spacetime interval is invariant in the flat, pseudo-Euclidean spacetime of STR; but, in general, in curved spacetime or in flat spacetime with a non-Euclidean topology (e.g. a plane wrapped around a cylinder), only the infinitesimal interval between the neighbouring points is invariant.

11 In practice the forces involved would be tremendous if the accelerations involved are too sudden. Not only would most living creatures fail to survive any too rapid change of direction, but physical objects like clocks would also be ripped apart. Generally, physicists overcome this 'technical' problem by speaking of ideal standard clocks, whose construction would not be so affected. STR may then be used to predict what would happen. We could of course slow the change of direction to a point where destructive forces would not ruin the experiment. This is exactly what was done in the Halfele–Keating experiment mentioned in the introduction to this chapter. But, if the relative speeds are too low, as in the Halfele–Keating experiment, then, although we will be able to measure differences between some kinds of physical clocks, the chances of being able to detect a difference in ageing in living creatures are indeed remote. So any measurable influence on the ageing of a traveller would require a high relative velocity. Although a round-trip to nearby stars is indeed possible in principle, such a journey would require a very large amount of fuel. Moreover, the rocket carrying a traveller on such a journey would require a tremendous amount of shielding to

protect the traveller from collisions with inter-stellar material. Hence, the chances of making such a trip are practically nil. See Taylor, E.F. and Wheeler, J.A. (1963) *Spacetime Physics* San Francisco: W.H. Freeman, especially the exercise on p. 174 and the exercise answer in the supplement, pp. 60–1.

12 This story of the twins has been simplified a good deal. For we must remember that the earth rotates on its axis as well as around the sun. And, since rotations are accelerations, giving rise to inertial forces, the twin on the Earth too must change frame, not just once, but all the time. But we could retell the story without problems and leading to the same essential result if we take the twin who stays 'behind' to be located in a frame of reference fixed by the background microwave radiation – a remnant from the very early universe – which is not rotating, according to the available observational evidence. This story has been told many, many times in almost as many ways. Further details may be found in almost all of these accounts; I have my own favourite versions – all of them generally clear and free from error: Bondi, H. (1967) *Assumption and Myth in Physical Theory* Cambridge: Cambridge University Press for a simple version; French, A.P. (1968) op. cit. and Taylor, E.F. and Wheeler, J.A. (1963) op. cit. for more detailed and mathematical accounts; and also, for a detailed philosophical analysis, Newton-Smith, W.H. (1980) *The Structure of Time* London: Routledge; finally, for an extremely thorough history of the various positions taken up on the paradox, Marder, L. (1971) *Time and the Space-Traveller* London: George Allen and Unwin.

13 This theme will be explored several times in this book – for it recurs in many of the different issues within the philosophy of space and time.

14 Strictly speaking, objects do not 'move along' world lines, for the world line is itself a representation of the object's motion; see Figure 4 (pp. 27–32).

15 The triplets involved in this thought experiment could not be genuine triplets, for some prior accelerations would be needed to set the two travelling triplets off on their inertial trajectories – if we make the assumption that at some time in the past all three and their clocks were together on Earth. Then we could not rule out those accelerations as the cause of the differences in ageing.

16 Although such synchronisations and checks would not be easy, they would not involve any synchronisation or check at any great distance. If the world lines come close to a genuine intersection, then, as the triplets and their clocks pass each other, a reasonably telling glimpse of each other and a fairly accurate reading of each other's clocks could be made. So long as the speeds involved and distances travelled in the experiment are sufficiently large – keeping the margin of error small in proportion to the results of the experiment – then there is likely to be a good match between theoretical predictions and actual observations.

17 More comprehensive details of this version of the paradox may be found in either Bondi, H. (1967) op. cit. or Newton-Smith, W.H. (1980) op. cit. We may also achieve the same general result with twins if they inhabit a universe which is cylindrical – one twin stays in the

same position (represented by a vertical line running up the side of the cylinder) and the other moves away inertially (represented by a straight line running around and up the cylinder to intercept the first world line after one complete circuit of the cylinder). No acceleration need be involved. Their spacetime journeys differ, so we have no reason to suppose that their clocks and ages will agree when they reunite.

18 Bohm, D. (1965) *The Special Theory of Relativity* New York: W.A. Benjamin; see especially pp. 165–67.
19 French, A.P. (1968) op. cit.
20 In general, GTR does provide a rather more elegant way of dealing with accelerated motion than STR.
21 Newton-Smith in his version admits that the clocks C_1 and C_3 will not agree at point R, but argues that there is nevertheless a reciprocal retardation between the two spacetime paths. The switch from C_2 to C_3 involves a discontinuity in which a continuous series of events in C_1's history is taken to occur simultaneously from the point of view of what he calls the composite clock consisting of C_2 and C_3 taken together. So the composite clock fails to record this time, which is shown to be equal to the difference in times on the clocks meeting at R. This strikes me as a sound explanation of why different spacetime paths do not in general agree, but the reciprocity involved here seems to do no more than clarify the differences between the two paths. For, if we admit the fact that the clocks meeting at R do in fact disagree, there is a real physical difference, a real asymmetry, which cannot be explained by forces *per se*, but only by spacetime geometry. See Newton-Smith, W.H. (1980) op. cit., pp. 192–5.

3 TRAVELLING LIGHT

1 This experiment, carried out in 1849, is described clearly in Hoffman, B. (1983) *Relativity and Its Roots* New York: Scientific American Books pp. 49–50.
2 Salmon, W.C. (1977) 'The philosophical significance of the one-way speed of light' *Nous* **11**, 3 pp. 253–92.
3 Ellis, G.F.R. and Williams, R.M. (1988) *Flat and Curved Space-Times* Oxford: Oxford University Press presents an excellent discussion of the physics involved.
4 Note that the conventionality of simultaneity is quite distinct from the relativity of simultaneity. The relativity of simultaneity arises from the fact that different observers viewing two events in the same frame from different perspectives will in general disagree about the temporal relations between those events. But if simultaneity is also conventional any given observer cannot be sure about which events in the same frame are simultaneous with some event local to the observer.
5 These are only three of the typical beliefs of one kind of scientific realist; there are other beliefs and there are other kinds of scientific realist; we shall be considering some of these in later chapters, especially Chapters 7, 9, and 10. For further details on the issues involved in scientific realism, see Harré, R. (1986) *Varieties of Realism*

Oxford: Basil Blackwell; Leplin, J. (ed.) (1984) *Scientific Realism* Berkeley: University of California Press; and Devitt, M. (1984) *Realism and Truth* Princeton: Princeton University Press.

6 See Quine, W.V.O. (1975) 'On empirically equivalent systems of the world' *Erkenntnis* **9** pp. 313–28.

7 Later in this section we shall examine another attempt to establish the one-way speed of light as an objective empirical fact: slow transport clock synchrony.

8 For an excellent account of various attempts to discover the one-way speed of light discussed here, see Hoffman, B. (1983) op. cit., ch. 4, as well as Salmon, W.C. (1977) op. cit. and Salmon, W.C. (1980) *Space, Time, and Motion* second edition Minneapolis, MN: University of Minnesota Press; Reichenbach H. (1957) *The Philosophy of Space and Time* New York: Dover also discusses a number of ways to determine the one-way speed of light and to thereby provide an objective definition of simultaneity.

9 Reichenbach (1957) op. cit., section 22.

10 See Grunbaum, A. (1969) 'Simultaneity by slow clock transport in the Special Theory of Relativity' *Philosophy of Science* **36** pp. 5ff.; Grunbaum's article is reprinted with an introduction co-authored by Salmon in Grunbaum, A. (1973) *Philosophical Problems of Space and Time* second edition Dordrecht, Holland: D. Reidel.

11 Winnie, J.A. (1970) 'Special relativity without one-way velocity assumptions', Parts I and II, *Philosophy of Science* **37** pp. 81–99, 223–38; see also Giannoni, C. (1978) 'Relativistic mechanics and electrodynamics without one-way velocity assumptions' *Philosophy of Science* **45** pp. 17–46.

12 Malament, D.B. (1977) 'Causal theories of time and the conventionality of simultaneity' *Nous* **11** pp. 293–300; see also Brown, H.R. (1990) 'Does the principle of relativity imply Winnie's (1970) equal passage times principle?' *Philosophy of Science* **57** pp. 313–24; and Friedman, M. (1983) *Foundations of Space-Time Theories* Princeton: Princeton University Press.

13 Salmon, W.C. (1980) op. cit., p. 119.

14 Salmon, W.C. (1980) op. cit., p. 120.

15 We should note that the sense of time travel here ought to be unobjectionable to those who oppose the idea of time travel on the grounds that it involves closed causal loops. No such loops are involved here, since all that is done is to 'skew' the plane of transmission in Newtonian spacetime.

16 Of course, the same problem would arise for STR if the spacetime structure were to be changed to allow the possibility of a similar kind of time travel.

17 Ellis, B. and Bowman, P.A. (1967) 'Conventionality in distant simultaneity' *Philosophy of Science* **34** pp. 116–36; and also Angel, R.B. (1980) *Relativity: the Theory and Its Philosophy* Oxford: Pergamon.

18 Descartes, R. (1988) *Selected Philosophical Writings* Cambridge: Cambridge University Press, translated by Cottingham, J., Stoothoff, R., and Murdoch, D.; see p. 58 and also p. 42.

19 As Hoffman, B. (1983) op. cit. points out, Rømer's calculation of the time taken is some 5 minutes short of the actual figure.
20 See, as well as Salmon (1980) op. cit., Grunbaum, A. (1969) op. cit., pp. 5ff.; his remarks in this paper are directed towards Ellis, B. and Bowman, P.A. (1967) op. cit.
21 See, for such a standard position on relativity, Hawking, S.W. and Ellis, G.F.R. (1973) *The Large Scale Structure of Space-Time* Cambridge: Cambridge University Press; of course, this standard view need not make the claim that nothing travels faster than light is actually a postulate of the theory: we may say that relativity is distinguished not by a belief in the fact that nothing may travel faster than light but by the assertion that the speed of light is an invariant; see Chapter 1.
22 See the excellent review and analysis in Recami, E. (1986) 'Classical tachyons and possible applications' *Nuovo Cimento* **9**, 6 pp. 1–178.
23 See, for details of the behaviour of tachyons in relativistic spacetimes, Friedman, M. (1983) *Foundations of Space-Time Theories* Princeton: Princeton University Press; and Recami, E. (1986) op. cit. Note also Earman, J. (1972) 'Causal propagation outside the null cone' *Australian Journal of Philosophy* **50** pp. 222–37; Earman raises a number of objections to the idea of spacelike travel, as we shall see in later sections of this chapter.
24 Tolman, R.C. (1917) *The Theory of the Relativity of Motion* Berkeley: University of California Press; and Bohm, D. (1965) *The Special Theory of Relativity* New York: Benjamin; see Recami, E. (1986) op. cit. for an analysis of this and other tales about tachyons, pp. 64–77.
25 Feynman, P.R. (1949) 'The theory of positrons' *Physical Review* **76** pp. 749–59.
26 Recami, E. (1986) op. cit., p. 66.
27 Elliot, R. (1981) 'How to travel faster than light' *Analysis* **41**, 1 pp. 4–6; see also Ray, C. (1982) 'Can we travel faster than light?' *Analysis* **42**, 1 pp. 50–2.
28 Elliot, R. (1981) op. cit., p. 5.
29 The problem of personal identity is a mare's nest which we shall leave for other braver souls; here we shall be as cavalier as Elliot and make some 'courageous' assumptions about what constitutes a person's identity; but see Parfit, D. (1984) *Reasons and Persons* Oxford: Oxford University Press for a more considered account of identity – he includes a detailed discussion of the relationship between identity and psychological continuity, pp. 219ff.

4 A CONVENTIONAL WORLD?

1 See Nerlich, G. (1976) *The Shape of Space* Cambridge: Cambridge University Press for a comprehensive study of Euclidean and non-Euclidean geometries; Nerlich reminds us that Euclidean geometry is constructed from these building-blocks: 'definitions of the various geometrical figures . . . axioms supposed to be too obvious to admit of proof . . . and postulates which admit of no proof from simpler statements, but which are not taken as obvious'; p. 51.

2 Poincaré, H. (1952) *Science and Hypothesis* New York: Dover, a translation by Greenstreet, W.J., of the original French edition of 1902.
3 Reichenbach, H. (1957) *The Philosophy of Space and Time* New York: Dover; this is a translation of Reichenbach's (1927) *Philosophie der Raum-Zeit-Lehre* by Maria Reichenbach, his wife.
4 The idea that two empirically equivalent theories are essentially the same is also reflected in van Fraassen, B.C. (1980) *The Scientific Image* Oxford: Oxford University Press.
5 See the articles by Schwartz and Linde in Hawking, S.W. and Israel, W. (eds) (1987) *300 Years of Gravitation* Cambridge: Cambridge University Press.
6 We should note that there are four distinct but inter-related levels of structure for space and time: metrical geometry, affine and conformal structures, and topological structure; see Figure 11 (p. 54).
7 See the accounts in Reichenbach, H. (1957) op. cit.; and Stewart, I. (1987) *The Problems of Mathematics* Oxford: Oxford University Press; Gauss is said to have discovered the essential ideas before the joint but independent 'discovery' by the other two, but inexplicably Gauss did not publish his results.
8 Riemann, G. (1873) 'On the hypotheses that lie at the foundations of geometry' *Nature* **8** p. 14; Riemann's (doctoral) paper of 1854 was translated by Clifford, W.K., the Cambridge mathematician responsible for the prophetic suggestion that matter might be no more than hill-like ripples in an otherwise flat space.
9 Minkowski, H. (1908) 'Space and time' reproduced in translation in Einstein, A., Lorentz, H.A., Weyl, H., and Minkowski, H. (1923) *The Principle of Relativity* London: Methuen. Although the spacetime of STR is not Euclidean *per se*, its global flatness, guaranteed by the absence of gravitating matter, gives it a Euclidean character. So STR spacetimes are often called 'pseudo-Euclidean'.
10 Reported in Calder, N. (1979) *Einstein's Universe* London: BBC Books.
11 Eddington, A.S. (1920) *Space, Time, and Gravitation* Cambridge: Cambridge University Press; ch. 7 gives a full report of the confirmation.
12 Reichenbach, H. (1957) op. cit.; Reichenbach's idea is similar in many respects to an earlier suggestion by Poincaré. See Sklar, L. (1974) *Space, Time, and Spacetime* Berkeley, CA: University of California Press for a detailed account of Poincaré's original suggestion. The basic idea of Reichenbach's approach is given here. In order to help us minimise problems of visualisation, Reichenbach asks us to concentrate on a two-dimensional world. He constructs a world in which two isolated communities exist on the two opposite surfaces of a glass block. Each member of the two groups is imagined to be a two-dimensional creature stuck to his or her particular surface. So no one can 'stand up' from the surface and take a privileged three-dimensional look around in order to determine the geometry of the world they inhabit. However, each group may carry out a series of measurements of distances on the surface itself using rigid metre rules and other measuring instruments. By doing so they may determine the metrical structure of the two-

dimensional surface upon which they move. Each group is able to view the measurements made on the other surface since the glass block is transparent. If there is any curvature on either surface, then we would expect the measurements taken to show this to be the case – just as measurements made on the surface of a spherical or an almost spherical object like the Earth can demonstrate its overall shape: on such a surface, for example, the internal angles of a triangle would be greater than 180 degrees. The 'top' surface is flat except for a central hemispherical 'hump'. The 'bottom' surface is flat everywhere. We would therefore expect the first group to say that their world is generally flat and Euclidean except for a central region which is curved. And the second group should say that their world is Euclidean everywhere. However, the observations of the second group agree perfectly with those of the first! Indeed, a flood of light shining uniformly downwards upon the world, casts shadows of the metre rulers on the top surface on to the bottom surface and these shadows coincide exactly with all lengths on the bottom. Indeed, everything, including the people on the lower surface, is affected in the same way.

13 Reichenbach calls such forces 'universal' since they affect different kinds of objects in exactly the same way. He notes that the effects of universal forces cannot be shielded. He distinguishes them from 'differential' forces which do not affect all objects in the same way: for example, one differential force is a magnetic force which has a greater effect on some materials than on others. See Reichenbach, H. (1957) op. cit., pp. 24–8.

14 Reichenbach, H. (1957) op. cit., pp. 43–4.

15 The problem of visualisation of geometries here is complicated by the fact that the geometries incorporated in modern gravitational theories are four-dimensional. An imagination conditioned to think of geometry spatially will find even a Euclidean four-dimensional view perplexing. For further details of such problems see Nerlich on Kant. The German philosopher Kant, writing in the eighteenth century, suggests that we must perceive the world spatio-temporally – in an important sense we construct the world in space and time; and he also claims that this construction must be Euclidean. He admits that our perception of the world as such implies nothing about the world as it really is – just that we must see it as such. See Nerlich, G. (1976) op. cit. for a full discussion of Kant's thinking. Also note the excellent review of various ways of visualising spatio-temporal worlds in Chandler, M. (1990) 'Philosophy of gravity: intuitions of four-dimensional curved spacetime' in Herget, D. (ed.) *More History and Philosophy of Science in Science Teaching* Tallahassee: Florida State University.

16 Noting, of course, that the idea of a straight line is defined by the affine structure and the idea of an angle is defined by the conformal structure; both of these structures are more general than the metrical structure so that different metrical geometries may be consistent with more than one of these other structures. See the discussion in Friedman, M. (1983) *Foundations of Space-Time Theories* Princeton, NJ: Princeton University Press.

17 A spacetime in which closed paths may be shrunk down continuously to a point is called 'simply connected'.
18 The topological discussion in this section is based on Reichenbach's worries about topology in Reichenbach, H. (1957) op. cit.; Reichenbach concentrates on the empirical equivalence between the plane and the surface of a torus to fuel his discussion; the example of the cylinder world and its covering space, the strip world, used in this section, may be found in Sklar, L. (1974) op. cit., ch. 2.
19 This dispute, raised by Newton-Smith, W.H. (1980) *The Structure of Time* London: Routledge, is discussed in Chapter 1.
20 The sequence of strips on the plane represents a 'covering space' for the cylinder – we may easily imagine the strips being wrapped around the cylinder so that each strip matches up with the events on the cylinder exactly.
21 Reichenbach (1957) op. cit., pp. 58-81.
22 Ehrenfest, P. (1917) 'In what way does it become manifest in the fundamental laws of physics that space has three dimensions?' *Proceedings of the Academy of Netherlands* **20** p. 200; and Weyl, H. (1922) *Space, Time, and Matter* New York: Dover: see p. 284.
23 See Davies, P.C.W. and Brown, J. (eds) (1988) *Superstrings: a Theory of Everything?* Cambridge: Cambridge University Press; the description here is based on Schwarz's own characterisation of superstring theory, which he gives in an interview reproduced in the book.
24 See Davies, P.C.W. and Brown, J. (eds) (1988) op. cit., pp. 192-210 for the complete text of the interview with Feynman.
25 See Linde, A. (1987) 'Inflation and quantum cosmology' in Hawking, S.W. and Israel, W. (eds) (1987) op. cit.
26 Barrow, J.D. (1983) 'Dimensionality' in McCrea, W.H. and Rees, M.J. (eds) (1983) *The Constraints of Physics* London: Royal Society pp. 337–46.
27 Note also that open spacetimes with a plane structure are non-compact and closed spacetimes with a spherical structure are compact; see Wald, R.M. (1984) *General Relativity* Chicago: Chicago University Press for technical details; a less daunting discussion may be found in Sklar (1974) op. cit.
28 See, for a particularly clear 'verdict' on this issue, Gribbin, J. and Rees, M. (1989) *Cosmic Coincidences* London: Bantam.
29 See Penrose, R. (1979) 'Singularities and time-asymmetry' in Hawking, S.W. and Israel, W. (eds) *General Relativity: an Einstein Centenary Survey* Cambridge: Cambridge University Press.
30 A closed universe is likely to have a 'spherically closed' geometry and an open universe is likely to have a flat or nearly flat geometry – given the available evidence; Ellis, G.F.R. and Williams, R.M. (1988) *Flat and Curved Space-Times* Oxford: Oxford University Press note that the properties of homogeneity and isotropy may be neatly explained in terms of the 'multiple image' torus universe – the small universe – in which the same finite collection of galaxies is viewed wherever we look over and over again: hence the 'appearance' of an infinite world in which everywhere and every direction seems to be the same – it looks the same because it is the same!

31 Ellis, G.F.R. (1978) 'Is the universe expanding?' *General Relativity and Gravitation* **8** pp. 87–94.
32 See Thurston, W.P. and Weekes, J.R. 'The mathematics of three-dimensional manifolds' *Scientific American* July 1984, and also Ellis, G.F.R. and Williams, R.M. (1988) op. cit.
33 Ellis, G.F.R. and Williams, R.M. (1988) op. cit., pp. 287–8.
34 In the small universe, there can be no surprises, for we can see all its history at least in principle; but the infinite, plane world could always throw up a surprise. For, in a genuinely infinite world, we cannot rule out the arrival of something which might disturb our environment from some part of the universe to which we have had no access so far. Our information about the past is incomplete, and so prediction would always carry a degree of uncertainty even if we were sure about the laws governing the universe. In a torus world, every prediction could be spot-on (at least in principle). But the possibility of differences in predictive power suggests that there may be some situation which allows us to distinguish the plane and the small universes on empirical grounds.
35 Duhem, P. (1954) *The Aim and Structure of Physical Theory* Princeton: Princeton University Press.
36 Quine, W.V.O. (1980) *From a Logical Point of View* second, revised edition Cambridge, MA: Harvard University Press. Note the astonishing extension of Duhem's basic idea here: from talk of the statements of science and our general experience to all statements, including analytic statements such as the laws of logic. We may indeed be thoroughly surprised by Quine's contention that the laws of logic and other analytic statements could be false.
37 Popper, K. (1963) *Conjectures and Refutations* London: Routledge p. 37; and Feynman, R. (1985) *Surely You're Joking, Mr. Feynman* London: Unwin Hyman pp. 338ff.
38 Quine, W.V.O. (1975) 'On empirically equivalent systems of the world', *Erkenntnis* **9** pp. 313–28.
39 This thesis is based on the weak statement of the Duhem–Quine thesis in Hesse, M.B. (1980) 'The hunt for scientific reason' in *Proceedings of the Conference of the Philosophy of Science Association 1980, vol. II*, Asquith, P. and Giere, R. (eds) East Lansing, MI: Michigan State University Press; the major problems for this thesis which follow and the restatement of a stronger version are all suggested by Hesse in this article.
40 If we drop this assumption then we might still take an anti-realist position by arguing that all equivalent theories are merely expressions of the same theory but that no alternative is the 'true' theory since each theory is reducible to exactly the same observational basis; that is, we would be denying the 'reality' of all 'theoretical' terms. This is essentially the position taken by Reichenbach: see the comments on version D of the thesis.
41 We would still need to be aware of the complexities involved in citing simplicity or some such other pragmatic reason for our choice; note, for example, the fact that one theory may be simpler than another

ontologically, but might nevertheless be more complex epistemologically: see Fine, A. (1986) *The Shaky Game* Chicago: Chicago University Press. See also the brief discussion of different kinds of simplicity in Ray, C. (1987) *The Evolution of Relativity* Bristol: Adam Hilger.
42 This is a vexed question. See, for example, the discussion in Sklar, L. (1974) op. cit., who also provides an extremely clear account of the issues under consideration in this chapter. The alternatives mentioned are both suggested by Sklar as possible routes through this epistemological minefield. But see Hesse, M.B. op. cit. and also Hesse, M.B. (1974) *The Structure of Scientific Inference* London: MacMillan for an argument in favour of taking simplicity as the touchstone of theory choice.
43 Hanson, N.R. (1958) *Patterns of Discovery* Cambridge: Cambridge University Press and Kuhn, T.S. (1970) *The Structure of Scientific Revolutions* second edition Chicago: Chicago University Press.
44 I shall explore the various arguments for and against the distinction in Chapter 6.
45 See Hesse, M.B. (1974) for the details of this argument.
46 Papineau, D. (1979) *Theory and Meaning* Oxford: Oxford University Press and Hookway, C. (1988) *Quine* Cambridge: Polity Press both give clear accounts of the difficulties involved.
47 See Friedman, M. (1983) op. cit., pp. 320ff.
48 Occam's razor (sometimes Ockham's razor) is named after William of Ockham's principle of parsimony or simplicity. Ockham, writing in England in the fourteenth century, frequently used this principle in his writings on logic and theology; its best known form is perhaps: you should not multiply entities without necessity, i.e. keep your metaphysics as simple as possible.

5 NEWTON AND THE REALITY OF SPACE AND TIME

1 For a comprehensive and definitive survey of this debate, see the excellent account by Barbour, J.B. (1989) *Absolute or Relative Motion?* vol. I Cambridge: Cambridge University Press.
2 Although I am sympathetic up to a point with anyone who might say that the *Principia* does not justify attributing the 'container' idea directly to Newton, I would justify the use of this idea in the Newtonian world-view, given his belief that God created the material world and placed it in space at a given time. Yes, Newton did resist the idea of space as a background 'substratum'; but this resistance derives from his displeasure with Descartes's notion that space is the primary substance and that matter is merely an aspect of this substance. As mentioned later in this chapter, the main target of Newton's arguments about space seems to have been Descartes.
3 Steps 1 to 8 follow the argument in the *Principia* (Scholium to definition 8) closely, but steps 9 and 10 are natural inferences from the substance and context of the argument; see Ray, C. (1987) *The Evolution of Relativity* Bristol: Adam Hilger ch. 1.
4 The reason for Newton's concentration on space rather than time and

for his failure to provide a clear argument for absolute space may be due to the fact that his target in the Scholium to the *Principia* was not Leibniz, who regarded both space and time as mere appearances with no absolute character, but Descartes, who had little to say about time; see Stein, H. (1967) 'On Newtonian spacetime' *Texas Quarterly* **10** pp. 174–200 and Ray, C. (1987) op. cit., pp. 10–11.

5 The counterfactual statement that 'if Venus had a moon, then it would have a roughly circular orbit' is given support by Newton's laws of motion; but the factual statement that 'all Presidents of the USA are male' is not given support by any law: it is supported only by the accidentally true generalisation that all Presidents to date have been male. For a full discussion of counterfactuals and laws see Armstrong, D.M. (1983) *What is a Law of Nature?* Cambridge: Cambridge University Press.

6 This definition is given in Earman, J.S. (1989) *World Enough and Space-Time* Cambridge, MA: MIT Press p. 12.

7 This leads to a notorious problem, which I shall not consider here: that of Newton's reluctance to accept the idea of action at a distance; for a full discussion see Koyre, A. (1957) *From the Closed World to the Infinite Universe* Baltimore: Johns Hopkins Press.

8 From Quaery 31 of Newton's *Optics*; see Hall, A.R. and Hall, M.B. (eds) (1962) *Unpublished Scientific Papers of Isaac Newton* Cambridge: Cambridge University Press p. 192. This collection also includes Newton's 'De gravitatione' – which might be regarded as a draft for the ideas in the Scholium. See the illuminating discussion of Newton's progress from 'De gravitatione' to the Scholium in Barbour, J.B. (1989) op. cit., pp. 609–39.

9 See Locke's essay in Ayer, A.J. and Winch, R. (eds) (1952) *British Empirical Philosophers* London: Routledge pp. 57ff.: or any comprehensive edition of Locke's works.

10 Locke, J. (1988) *Essays on the Laws of Nature* (edited by von Leyden, W.) Oxford: Oxford University Press pp. 258–9.

11 See Ayer, A.J. and Winch, R. (eds) (1952) op. cit., pp. 69–78.

12 Alexander, H.G. (ed.) (1956) *The Leibniz–Clarke Correspondence* Manchester: Manchester University Press p. 71.

13 Leibniz's third letter, paragraph 4: in Alexander, H.G. (ed.) (1956) op. cit., pp. 25–6.

14 One of many versions of the principle quoted in Mates, B. (1986) *The Philosophy of Leibniz* Oxford: Oxford University Press p. 152.

15 Mates, B. (1986) op. cit., p. 132.

16 The rule of *modus tollens*, 'if P then Q; not Q; therefore not P', is a valid deduction.

17 We may distinguish between two versions of PSR: causal and theological; as Earman points out, we need more than a causal reading of PSR to drive home Leibniz's argument. See Earman, J.S. (1989) op. cit., p. 118. Teller argues that PII might be sufficient to drive home Leibniz's point: if the inhabitants of X have no way of distinguishing X from Y even in principle, then X and Y must be identical – hence the notions of different spatial positions and of different points in time

mean nothing by themselves; see Teller, P. (1987) 'Space-time as a physical quantity' in Achinstein, P. and Kagon, R. (eds) *Kelvin's Baltimore Lectures and Modern Theoretical Physics* Cambridge, MA: MIT Press; and also Teller, P. (1991) 'Substance, relations, and arguments about the nature of space-time' *Philosophical Review* forthcoming, which gives a summary of Teller's position on PII.

18 Alexander, H.G. (ed.) (1956) op. cit., pp. 45–6.
19 See Cleomedes' reports of the Stoic arguments in Sorabji, R. (1988) *Matter, Space, and Motion* London: Duckworth.
20 Alexander, H.G. (ed.) (1956) op. cit., p. 48.
21 However, we might be inclined to regard Leibniz's idea of *vis viva*, discussed later in this section, as a suggestion that Leibniz would not have been altogether antagonistic to Sklar's notion of motion as a brute fact.
22 John Earman identifies two main aspects to the plea for relationism: (a) space and time do not have elements of any kind which allow us to talk in terms of absolute motion; and (b) spatio-temporal relations amongst objects and events are not dependent upon any underlying 'substratum' of spatio-temporal points. I agree with Earman when he says that (a) entails (b), for (a) is clearly more general than (b). He goes on to say that many commentators maintain that Newton and his supporters, presumably including Clarke, argue that: because (b) is mistaken and there are spatial points or regions and temporal moments, (a) must also be wrong. We can see from the reconstruction of Clarke's argument, that Earman is right to challenge this view of the Newtonian argument. There is no one-way entailment in Clarke's argument. Assumption 2 provides a two-way link between points/regions and absolute motion; and assumption 3 provides a similar link between absolute motion and inertial forces. The three basic ideas here are being inter-defined. But, when we look at what gets the argument off the ground, inertial forces take the lead. In the original argument in the Scholium, Newton sticks at proving the case for absolute motion in terms of inertial forces. As we have already noted no further link is made between motion and the parts of space. But Clarke is faced with a challenge to the link between motion and the parts of space, given Leibniz's assertion that space cannot have distinctive parts in any way.
23 A useful summary of this correspondence appears in Howard Stein's 'Some philosophical prehistory of General Relativity' in Earman, J.S., Glymour, C.N., and Stachel, J.J. (eds) (1977) *Foundations of Space-Time Theories* Minneapolis: University of Minnesota Press.
24 Perhaps the reason for the lack of clarity is Clarke's occasional tendency to depart from the Newtonian standpoint and regard space and time as attributes: see Alexander, H.G. (ed.) (1956) op. cit., p. xxviii.
25 Notice that Newton's argument does not rely on any particular conceptions of God and of how God interacts with the world.
26 Alexander, H.G. (ed.) (1956) op. cit., p. 74.
27 Alexander, H.G. (ed.) (1956) op. cit., p. 74.
28 *Vis viva* is identified by Leibniz as the quantity mv^2 – a quantity which

he believes to be conserved during collisions between objects; for a discussion of *vis viva* and its relation to the ideas of momentum and kinetic energy see Iltis, C. (1970) 'Leibniz and the *vis viva* controversy' *Isis* **63** pp. 26ff.: and Wilson, M. (1976) 'Leibniz's dynamics and contingency in nature' in Machamer, P.K. and Turnbull, R.G. (eds) *Motion and Time, Space and Matter* Columbus, OH: Ohio State University Press.

29 One interesting question, arising from Leibniz's relationism, asks whether we should consider only relations between actual objects or events, or we should take relationism to involve possible relations as well. For details of this issue see Sklar, L. (1974) *Space, Time, and Spacetime* Berkeley: University of California Press for a general review of the problems connected with possible relations; Manders, K. (1982) 'On the spacetime ontology of physical theories' *Philosophy of Science* **49** pp. 575–90; and Mundy, B. (1989) 'On quantitative relationist theories' *Philosophy of Science* **56** pp. 582–600.

30 See Graves, J.C. (1971) *The Conceptual Foundations of Contemporary Relativity Theory* Cambridge, MA: MIT Press.

31 Given Newton's unhappiness with the idea of action at a distance, there is some evidence to suggest that he thought of forces as mediated by spirit in the absence of any substantial material ether; see Hall, A.R. and Hall, M.B. (eds) (1962) op. cit.

32 See Stein, H. (1967) op. cit.

33 Sklar, L. (1974) op. cit., pp. 225ff.

34 Einstein understood this problem and said that his own dynamical conception of spacetime dissolved the inconsistency inherent in Newton's account.

35 Friedman, M. (1983) *Foundations of Space-Time Theories* Princeton: Princeton University Press; see also Friedman, M. (1982) 'Theoretical explanation' in Healey, R.A. (ed.) *Reduction, Time, and Reality* Cambridge: Cambridge University Press.

36 Earman, J.S. (1989) op. cit., see pp. 4, 154, and 170–4.

6 MACH AND THE MATERIAL WORLD

1 Here, I disagree with those like John Gribbin who say that Mach did not improve on Berkeley's argument; see Gribbin, J. (1984) 'The bishop, the bucket, Newton and the universe' *New Scientist* **104** 1435/36 pp. 12–16.

2 Mach, E. (1911) *The Conservation of Energy* La Salle, IL: Open Court pp. 91, 87.

3 Mach, E. (1960) *The Science of Mechanics* sixth edition La Salle, IL: Open Court p. xxiii and Mach, E. (1943) *Popular Scientific Lectures* fifth edition La Salle, IL: Open Court p. 207. (Original publication dates of German editions 1883 and 1895 respectively.)

4 The least sympathetic of recent writers is probably John Earman; see Earman, J.S. (1970) 'Who's afraid of absolute space?' *Australasian Journal of Philosophy* **48** pp. 287–319; and Earman, J.S. (1989) *World Enough and Space-Time* Cambridge, MA: MIT Press.

5 See the repeated references to Mach's lasting influence in Einstein, A. (1969) 'Autobiographical notes', especially pp. 21, 29, 53, 63, and 67, in Schilpp, P. (ed.) *Albert Einstein: Philosopher-Scientist* La Salle, IL: Open Court pp. 2–95.
6 See, for example, Brans, C. and Dicke, R.H. (1961) 'Mach's Principle and a relativistic theory of gravitation' *Physical Review* **124** pp. 925–35; Sciama, D.W., Waylen, P.C., and Gilman, R.C. (1969) 'Generally covariant integral formulations of Einstein's field equations' *Physical Review* **187** pp. 1762–84; Raine, D.J. (1981) *The Isotropic Universe* Bristol: Adam Hilger; and Barbour, J.B. and Bertotti, B. (1982) 'Mach's Principle and the structure of dynamical theories' *Proceedings of the Royal Society* **382** pp. 295–306.
7 See, for a history and appraisal of this movement, Hanfling, O. (1981) *Logical Positivism* Oxford: Basil Blackwell. For a display of Mach's influence on a philosopher of science, see Reichenbach, H. (1957) *The Philosophy of Space and Time* New York: Dover. As logical positivism developed, its proponents realised that Mach's own brand of positivism was inadequate for their purposes; see Carnap, R. (1934) 'Protocol statements and the formal mode of speech' in Hanfling, O. (ed.) (1981) *Essential Readings in Logical Positivism* Oxford: Basil Blackwell. But, in the early days of logical positivism, Mach was very much regarded as a guiding light. Indeed, the Vienna Circle of logical positivists was originally known as the Ernst Mach Society.
8 Mach, E. (1960) op. cit., p. 280.
9 Mach, E. (1960) op. cit., p. 286.
10 Mach, E. (1960) op. cit., p. 591.
11 Mach, E. (1960) op. cit., pp. 578–9.
12 Mach, E. (1943) op. cit., p. 192.
13 Mach, E. (1960) op. cit., p. xxiii.
14 Ray, C. (1987) *The Evolution of Relativity* Bristol: Adam Hilger gives an example of calculating the area under a curve: if we use Simpson's rule or the trapezoidal method to calculate the area we would be fulfilling the demand for economy of labour, but if we use Euler's method we would satisfy the demand for economy of form. Mach gives us no advice about how we should resolve such difficulties.
15 See Hesse, M.B. (1974) *The Structure of Scientific Inference* London: Macmillan p. 223; its lack of some such distinction prevents Mach's theory from giving us satisfactory and rigorous criteria of simplicity.
16 Friedman, M. (1983) *The Foundations of Space-Time Theories* Princeton: Princeton University Press, especially the final chapter; a less technical exposition of his ideas appears in Friedman, M. (1981) 'Theoretical explanation' in Healey, R.A. (ed.) *Reduction, Time, and Reality* Cambridge: Cambridge University Press; see also the discussion on Newton and Leibniz in the last section of Chapter 5 above.
17 The claims that observation is theory-laden and that there is no clear-cut distinction between observational and theoretical statements are also discussed in a straightforward way in Hanson, N.R. (1958) *Patterns of Discovery* Cambridge: Cambridge University Press; and Newton-Smith, W.H. (1981) *The Rationality of Science* London: Routledge. The

quotation given is from Kuhn, T.S. (1970) *The Structure of Scientific Revolutions* second edition Chicago: Chicago University Press pp. 111–12.

18 I might be accused of taking Kuhn's position to relativist extremes here – of constructing a straw man. In the Postscript to Kuhn, T.S. (1970) op. cit., Kuhn denies that he is a relativist. I take a relativist (about scientific truth) to be someone who believes that there is no way of deciding between alternative theories on the basis of their correspondence with the world. Some people might well be relativists about all aspects of science: not only might anything go as regards truth, any old methodology might be fine. Kuhn is not one of these: he does recognise the need for a coherent and structured methodology to control the development of science. However, science develops, for Kuhn, not by getting closer to the truth about the world, but by producing the conditions for successful puzzle-solving. Undoubtedly some theories will be better suited to solving puzzles than others. So Kuhn is certainly not a relativist as far as methodology and puzzle-solving capacities are concerned. But his message in Kuhn, T.S. (1970) op. cit. is relativist as far as truth is concerned. He says (p. 206): 'there is, I think, no theory-independent way to reconstruct phrases like "really there"; the notion of a match between the ontology of a theory and its "real" counterpart in nature now seems to me to be illusive in principle'. My characterisation of Kuhn is flesh and blood, not straw.

19 See Maxwell, G. (1962) 'The ontological status of theoretical entities' in Feigl H. and Maxwell G. (eds) *Minnesota Studies in the Philosophy of Science* vol. III Minneapolis: University of Minnesota Press; the spectrum view is discussed in Newton-Smith, W.H. (1981) op. cit., ch. 2.

20 The phrase 'contingent infallibility or incorrigibility' is suggested by Zahar, E. (1980) 'Second thoughts about Machian positivism' *British Journal for the Philosophy of Science* **32** p. 268.

21 See Feyerabend, P.K. (1988) *Against Method* second edition London: Verso; and Feyerabend, P.K. (1981) *Philosophical Papers* vols I and II Cambridge: Cambridge University Press.

22 van Fraassen, B. (1980) *The Scientific Image* Oxford: Oxford University Press.

23 This first issue is fully discussed by Hacking, I. (1983) *Representing and Intervening* Cambridge: Cambridge University Press: see especially part B.

24 Hacking, I. (1983) op. cit., pp. 164–5.

25 This point is made by Peter Galison in Galison, P. (1988) 'Philosophy in the laboratory' *Journal of Philosophy* **85**, 10 pp. 525–7; see also Galison. P. (1987) *How Experiments End* Chicago: Chicago University Press; and Hacking, I. (1988) 'On the stability of the laboratory sciences' *Journal of Philosophy* **85**, 10 pp. 507–14.

26 Hacking, I. (1983) op. cit., p. 165.

27 See Putnam, H. (1984) 'What is realism?' in Leplin, J. (ed.) *Scientific Realism* Berkeley: University of California Press for details of this standard realist position; note also Leplin's clear and concise introduction to the issue of realism.

7 EINSTEIN AND ABSOLUTE SPACETIME

1 The clearest statement of Reichenbach's position appears in Reichenbach, H. (1924) 'The theory of motion according to Newton, Leibniz, and Huyghens' reprinted in Reichenbach, H. (1959) *Modern Philosophy of Science* London: Routledge; see, for an example of someone who follows Reichenbach's line, Alexander, H.G. (ed.) (1956) *The Leibniz–Clarke Correspondence* Manchester: Manchester University Press.
2 See, for example, Earman, J.S. (1970) 'Who's afraid of absolute space?' *Australasian Journal of Philosophy* **48** pp. 287–319 and Earman, J.S. (1989) *World Enough and Space-Time* Cambridge, MA: MIT Press; Sklar, L. (1974) *Space, Time, and Spacetime* Berkeley: University of California Press; Gardner, M.R. (1977) 'Relationism and relativity' *British Journal for the Philosophy of Science* **28** pp. 215–33; Friedman, M. (1983) *Foundations of Space-Time Theories* Princeton: Princeton University Press. The best general review of the philosophical background to GTR is probably Stein, H. (1977) 'Some philosophical prehistory of General Relativity' in Earman, J.S., Glymour, C.N., and Stachel, J.J. (eds) *Foundations of Space-Time Theories* Minneapolis: University of Minnesota Press.
3 The original field equations may be expressed in coordinate form in the following way:

$$R_{ik} - 1/2Rg_{ik} = 8\pi T_{ik}$$

where the metrical function on the left-hand side of the equation represents the geometry of spacetime, and the stress-energy function on the right-hand side represents the distribution of mass and energy. In its abstract geometrical formulation, the equation may be expressed as follows:

$$\mathbf{G} = 8\pi \mathbf{T}$$

See Misner, C.W., Thorne, K.S., and Wheeler, J.A. (1973) *Gravitation* New York: W.H. Freeman.
4 This feature of GTR's spacetime is linked with the ideas involved in Einstein's principle of equivalence. This principle was motivated by the established observational equivalence between inertial and gravitational masses and forces. In a strong form, the principle demands that those laws governing the motion of particles in STR should also hold good in the context of GTR. See Misner, C.W., Thorne, K.S., and Wheeler, J.A. (1973) op. cit., pp. 312–15. This is achieved by considering a sufficiently small 'local' portion of spacetime around the particle in which we are interested. In effect, the motion is then referred to a local inertial frame in which the particle moves in a straight line at a uniform speed. This idea is the key to GTR: Einstein moves away from the idea of a global to a local frame of reference for motion. Spacetime *may* be globally curved with motions being consequently complex. But *locally* we should treat the spacetime of GTR as if it were the 'flat' spacetime of STR with an elegant and simple description of all motions. The dynamics of motion may then be

referred to a local limiting inertial frame of reference. This mathematical and theoretical simplicity may only be achieved if we exclude any additional curvature terms from the field equations. For such a term blocks the possibility of 'flat' spacetime in the absence of matter – even locally curvature will be apparent. A consequence of the principle is that physical laws should be effectively the same in the contexts of STR and of the local inertial frames of GTR. However, there is no general way of preventing curvature terms from appearing at a local level in physical laws when we try to use them in GTR.

5 Misner, C.W., Thorne, K.S., and Wheeler, J.A. (1973) op. cit., pp. 312–15.
6 Einstein, A. (1918) 'Principles of general relativity' *Annalen Physik Leipzig* **55** pp. 241–5; see note at foot of p. 241.
7 Wheeler, J.A. (1964) 'Mach's Principle as a boundary condition for Einstein's equations' in Chiu, H.Y. and Hoffman, W.F. (eds) *Gravitation and Relativity* New York: Benjamin.
8 The determination must be 'unique', otherwise two or more spacetime structures may be compatible with the same mass–energy distribution; hence, there would be more to spacetime than its material contents.
9 These three senses of 'absolute' are based on Michael Friedman's illuminating analysis of the issue in Friedman, M. (1983) op. cit., pp. 62–70.
10 For details of spacetime accounts of Newtonian physics see Friedman, M. (1983) op. cit.; Stein, H. (1967) 'Newtonian spacetime' *Texas Quarterly* **10** pp. 174–200; Malament, D.B. (1986) 'Newtonian gravity, limits, and the geometry of space' in Colodny, R.G. (ed.) *From Quarks to Quasars: Philosophical Problems of Modern Physics* Pittsburgh: University of Pittsburgh Press; Cartan's classic attempt to rewrite Newtonian gravity in spacetime form is discussed in Misner, C.W., Thorne, K.S., and Wheeler, J.A. (1973) op. cit., pp. 291ff.
11 We should not regard the possibility of rewriting Newtonian theory in a modern spacetime form as an invitation to rewrite history and start imputing spacetime language to Newton himself, as Earman, J.S. (1989) op. cit. does occasionally; the example given here concerns Newton's claim in the Scholium that time flows and Earman's statement that Newton did not really mean this; see p. 8. Earman also accuses Newton of making a mistake when he condemns Newton for not being sharp enough to realise that he did not need Newtonian space and time as a basis for absolute acceleration; he could have employed 'neo-Newtonian' spacetime instead, with an inertial structure but no overall absolute Newtonian framework; see Earman, J.S. (1989) op. cit., p. 3. (Neo-Newtonian spacetime incorporates the Newtonian idea of simultaneity, but drops the notion of being at the same point at different times and therefore drops the notion of an overall 'rigid' framework in the Newtonian sense – informally one may think of neo-Newtonian spacetime as a half-way house between Newtonian ideas and the Minkowski spacetime of STR; see Sklar (1974) op. cit. for a full discussion of the properties of this spacetime.) But, historically Newton had no such option. And, conceptually, it is far from clear how Newton

might have utilised such a structure. See the review of Earman's book by Hoefer, C. and Ray, C. (1991) 'Earman: *World Enough and Space-Time*' *British Journal for the Philosophy of Science* **42** 3 for a detailed discussion of this and other related points arising from Earman's treatment of spacetime physics.

12 Note the warning on p. 725 of Misner, C.W., Thorne, K.S., and Wheeler, J.A. (1973) op. cit.: even if we completely determine all the local geometrical properties by, say, imposing a demand for homogeneity and isotropy, we do not thereby determine the global topology of the spacetime.

13 See Weinberg, S. (1972) *Gravitation and Cosmology* New York: Wiley; for a philosopher of physics who follows the lead given by Weinberg, see Sklar, L. (1974) op. cit.

14 This and other effectively 'empty' models of GTR are discussed in Hawking, S.W. and Ellis, G.F.R. (1973) *The Large Scale Structure of Space-Time* Cambridge: Cambridge University Press pp. 117ff.

15 For details of the Kerr solution and Gödel's solution of the field equations see Hawking, S.W. and Ellis, G.F.R. (1973) op. cit., pp. 162–70; the Oszvàth and Schücking solution is discussed together with Gödel's solution in Friedman, M. (1983) op. cit., pp. 206–15.

16 Two of his most important papers, Friedmann, A. (1922) 'On the curvature of space' and Friedmann, A. (1924) 'On the possibility of a world with constant negative curvature', appear in translation in Bernstein, J. and Feinberg, G. (eds) (1986) *Cosmological Constants: Papers in Modern Cosmology* New York: Columbia University Press; see also Misner, C.W., Thorne, K.S., and Wheeler, J.A. (1973) op. cit. for full details of his models plus p. 751 for a biographical note.

17 See Rindler, W. (1977) *Essential Relativity* second edition Berlin: Springer p. 243.

18 Raine, D.J. (1975) 'Mach's Principle in General Relativity' *Monthly Notices of the Royal Astronomical Society* **171** pp. 507–28, and Raine, D.J. (1981) *The Isotropic Universe* Bristol: Adam Hilger; see also Sciama, D.W., Waylen, P.C., and Gilman, R.C. (1969) 'Generally covariant integral formulation of Einstein's field equations' *Physics Review* **187** pp. 1762–84.

19 Other attempts to construct Machian variants of gravitational theory include Brans, C. and Dicke, R.H. (1961) 'Mach's Principle and a relativistic theory of gravitation' *Physical Review* **124** pp. 925–35; Barbour, J.B. and Bertotti, B. (1977) 'Gravity and inertia in a Machian framework' *Nuovo Cimento* **38B** pp. 1–27; Barbour, J.B. and Bertotti, B. (1982) 'Mach's Principle and the structure of dynamical theories' *Proceedings of the Royal Society London* **382** pp. 295–306.

20 See Collins, C.B. and Hawking, S.W. (1973) 'Why is the universe isotropic?' *Astrophysics Journal* **180** pp. 317–34; and Will, C.W. (1981) *Theory and Experiment in Gravitational Physics* Cambridge: Cambridge University Press. Current estimates for the margin of error on the assertion that there is no overall rotation are less than 0.1%.

21 See Ray, C. (1987) *The Evolution of Relativity* Bristol: Adam Hilger pp. 111–15 for a discussion of laws and implicit ranges of application.

22 Bondi, H. (1967) *Assumption and Myth in Physical Theory* Cambridge: Cambridge University Press.
23 Hawking, S.W. and Ellis, G.F.R. (1973) op. cit., p. 117.
24 Of course, we might discover tachyons or some other material which breaks these conditions: but until we have some evidence for such strange kinds of matter there is little reason to allow solutions involving them except as mathematical curiosities.
25 See Chapter 9 for a more detailed discussion of the early universe and its evolution.
26 See, for the general background to Einstein's development of GTR, Pais, A. (1982) *Subtle is the Lord* Oxford: Oxford University Press; and Earman, J.S. and Glymour, C. (1978) 'Lost in tensors' *Studies in the History and Philosophy of Science* **9** pp. 251–78. For accounts of the hole argument see Butterfield, J. (1989) 'The hole story' *British Journal for the Philosophy of Science* **40** pp. 1–28; Earman, J.S. and Norton, J. (1987) 'What price space-time substantivalism? The hole story' *British Journal for the Philosophy of Science* **38** pp. 515–25; Earman, J.S. (1989) op. cit.; Teller, P. (1991) 'Substance, relations, and arguments about the nature of space-time' *Philosophical Review*.
27 In practice, the technicalities are such that we may only do this in highly symmetrical and simple spacetimes.
28 There are a number of complex issues involved in the possibility of such a deterministic picture: we may only construct such a global time-slice (a Cauchy hypersurface) in certain well-behaved spacetimes. We shall consider the complications, connected with the ideas of closed timelike loops, naked singularities, and topological holes, in Chapters 8 and 10.
29 See Misner, C.W., Thorne, K.S., and Wheeler, J.A. (1973) op. cit. for details of general covariance and its role in GTR; and Ray, C. (1987) op. cit., especially ch. 2, for an informal discussion of the idea and status of general covariance.
30 See Kretschmann, E. (1917) 'On the physical meaning of the relativity postulate' *Annalen Physik Leipzig* **53** pp. 574–614; and also Friedman, M. (1983) op. cit.
31 See Teller, P. (1991) op. cit.
32 This is a simplified account of the argument, omitting all technical details, which may be found in Earman, J.S. (1989) op. cit., pp. 175–80.
33 See Teller, P. (1991) op. cit. for a discussion of these approaches taken by, respectively, Maudlin, T. (1988) 'Substances and space-time: what Aristotle would have said to Einstein' (forthcoming) and Butterfield, J. (1989) op. cit.

8 TIME TRAVEL

1 Wells, H.G. (1958) *Selected Short Stories* London: Penguin.
2 Mellor, D.H. (1981) *Real Time* Cambridge: Cambridge University Press.
3 For amplification of this idea of change, applying to things but not

events, see Mellor, D.H. (1981) op. cit., p. 9 and pp. 119ff.
4 See the section 'Convention and topology' in Chapter 4.
5 See, for example, the discussion of the role of thermodynamics and entropy in questions of time asymmetry in Swinburne, R. (1981) *Space and Time* London: Macmillan; and Penrose's ideas on gravitational curvature and time asymmetry may be found in Penrose, R. (1979) 'Singularities and time-asymmetry' in Hawking, S.W. and Israel, W. (eds) (1979) *General Relativity* Cambridge: Cambridge University Press pp. 581–638; see also the extremely helpful general analyses in (from a philosophical perspective) Sklar, L. (1974) *Space, Time, and Spacetime* Berkeley, CA: University of California Press; and (from a physical perspective) Davies, P.C.W. (1974) *The Physics of Time Asymmetry* Guildford: Surrey University Press.
6 A given spacetime may, however, have more than one surface of simultaneity through a given point, e.g. Minkowski spacetime has two surfaces of simultaneity: the plane (for observers at rest relative to each other) and the hyperboloid (for observers in relative motion).
7 Dummett, M.A.E. (1954) 'Can an effect precede its cause?' *Proceedings of the Aristotelian Society, Supplementary Volume* **28** pp. 27–44; Dummett, M.A.E. (1964) 'Bringing about the past' *Philosophical Review* **69** pp. 497–504; Dummett, M.A.E. (1986) 'Causal loops' in Flood, R. and Lockwood, M. (eds) *The Nature of Time* Oxford: Basil Blackwell pp. 135–69.
8 Such manoeuvres are called 'bilking' strategies by Horwich, P. (1987) *Asymmetries in Time* Cambridge, MA: MIT Press; see pp. 91ff.
9 An interesting review of wormholes appears in Redmount, I. (1990) 'Wormholes, time travel and quantum gravity' *New Scientist* **126**, 1714 pp. 57–61; the information in Figure 28 is based on the excellent graphics in this article.
10 Feynman, P.R. (1949) 'The theory of positrons' *Physical Review* **76** pp. 749–59; see also Horwich, P. (1987) op. cit. for a clear analysis of the implications of Feynman's account of backwards causation.
11 Mellor, D.H. (1981) op. cit.
12 It may be possible to construct situations in which travel backwards in time does not allow the construction of closed causal loops. For example, the skewed Newtonian spacetime discussed in Chapter 4 in which light travels forwards in time when moving in one spatial direction, but backwards in time when moving in the opposite spatial direction: in such a case the spacetime is constructed to give a limited sense to the notion of backwards travel – it is backwards relative to a notional plane of 'instantaneous' simultaneity; but the specifications of this spacetime guarantee that nothing can travel along a world line with a tilt less than that defined by the skewed plane of simultaneity as defined by light signals; hence, there can be no sense in talking of closed loops here.
13 There are other, more complex, spacetimes in which closed non-spacelike causal curves may be constructed, for example: Gödel's universe and Taub-NUT spacetimes; for details of these causally unstable spacetimes see Hawking, S.W. and Ellis, G.F.R. (1973) *The*

Large-Scale Structure of Space-Time Cambridge: Cambridge University Press, ch. 5; for a less technical discussion of Gödel's universe, see either Horwich, P. (1987) op. cit. or Sklar, L. (1974) op. cit.; further philosophical analysis of the technical details is given in Malament, D. (1984) 'Time travel in the Gödel universe' in Asquith, P.D. and Kitcher, P. (1985) *Proceedings of the Philosophy of Science Association* vol. II Minneapolis: University of Minnesota Press.

14 This is based on a similar situation characterised in Earman, J. (1972) 'Causal propagation outside the null cone' *Australasian Journal of Philosophy* **50** pp. 222–37.

15 See Lewis, D. (1976) 'The paradoxes of time travel' *American Philosophical Quarterly* **13** pp. 145–52, for a classic discussion of this kind of attack on time travel.

16 See Dummett, M.A.E. (1986) op. cit.

17 Harrison, J. (1979) 'Analysis problem number 18' *Analysis* **39** pp. 65–6; see also Harrison, J. (1980) 'Report on *Analysis* problem number 18' *Analysis* **40** pp. 65–9.

18 Malament, D. (1985) 'Minimal acceleration requirements for time travel in Gödel space-time' *Journal of Mathematical Physics* **26** pp. 774–7.

19 Horwich, P. (1975) 'On some alleged paradoxes of time travel' *Journal of Philosophy* **72** pp. 432–44, and Horwich, P. (1987) op. cit., pp. 111–28.

20 I remain uncertain about the strength of Horwich's point: I have argued in its favour in Ray, C. (1987) *The Evolution of Relativity* Bristol: Adam Hilger pp. 127–9; but I suppose I am now less than satisfied with brute facts than others might be.

21 This story about wishing and letters is based on the examples given by Dummett (1964) op. cit. and Mellor (1981) op. cit. I have tried to capture Dummett's connection between wishing that something had been the case (and acting on this wish) and it having been the case, something which is absent in Mellor's example; but also to avoid the associations of Dummett's 'dancing chief' example, which Mellor quite rightly tries to avoid.

22 Riggs, P.J. (1991) 'A critique of Mellor's argument against backwards causation' *British Journal for the Philosophy of Science* **42** (forthcoming) argues that Mellor's views imply an acceptance of any account in terms of forwards causation 'regardless of how tenuous and improbable such an explanation might be'. But this point of view seems to neglect the tremendous difficulties involved in maintaining backwards-type causal explanations. I do not think we could win whatever we were to decide. But fortunately there seems little need to make a decision, given the total absence of empirical evidence for backwards causation.

9 EINSTEIN'S GREATEST MISTAKE?

1 de Sitter, W. (1917) 'On Einstein's theory of gravitation and its astronomical consequences' *Monthly Notices of the Royal Astronomical Society* **78** pp. 3–28; for more recent attacks on the cosmological constant see: Gamov, G. (1957) 'Modern cosmology' in Munitz, M.

(ed.) (1957) *Theories of the Universe* New York: Free Press; Pais, A. (1982) *Subtle is the Lord* Oxford: Oxford University Press; Hawking, S.W. (1982) 'The cosmological constant and the weak anthropic principle' in Duff, M.J. and Isham, C.J. (eds) *Quantum Structure of Space and Time* Cambridge: Cambridge University Press. See also the excellent history of early twentieth-century cosmology, which focuses on de Sitter and Einstein's debates about cosmology: Kerszberg, P. (1989) *The Invented Universe* Oxford: Oxford University Press. Another history of cosmology which discusses the issues raised in this chapter, more wide-ranging than Kerszberg's, is North, J.D. (1965) *The Measure of the Universe* Oxford: Oxford University Press; see Chapter 2 of North for problems associated with the Newtonian account.
2 Einstein, A. (1931) 'Zum kosmologischen Problem der allgemeinen Relativitatstheorie' *Sitzungsberichte der Preussischen Akademie der Wissenschaften* **142** pp. 235–7.
3 Einstein, A. (1917) 'Cosmological considerations on the general theory of relativity' in Einstein, A., Lorentz, H.A., Weyl, H., and Minkowski, H. (1923) *The Principle of Relativity* London: Methuen.
4 Newton, I. (1961) *Correspondence* vol. III Cambridge: Cambridge University Press.
5 Neumann, C. (1896) *Uber das Newton'sche Prinzip der Fernwirkung* Leipzig: B.G. Teubner, see pp. 165ff.: see also Seeliger, H. (1895) 'Uber das Newton'sche Gravitationsgesetz' *Sitzungsberichte der Bayerischen Akademie der Wissenschaften* **26** pp. 373–90 and Seeliger, H. (1898) 'On Newton's law of gravitation' *Popular Astronomy* 5 pp. 544–51. Einstein acknowledges his debt to Seeliger in Einstein, A. (1954) *Relativity: the Special and the General Theory* London: Methuen: this is a translation of a popular account written by Einstein in 1917.
6 Newton, I. (1729) *Principles of Natural Philosophy* London: Dawsons.
7 Einstein, A. (1918) 'Principles of general relativity' *Annalen Physik Leipzig* **55** pp. 241–4; note also Mach, E. (1883) *The Science of Mechanics* sixth edition (1960) La Salle, IL: Open Court, p. 207.
8 Pais, A. (1982) op. cit.
9 See Einstein, A. (1917) op. cit.
10 de Sitter, W. (1917) 'On the relativity of inertia' *Proceedings of the Academy of the Netherlands* **19** pp. 1217–25.
11 Ernst Mach, as we have seen, continually reminded his readers, including Einstein, of the need for economy in our descriptions of the physical world. He says that the 'fundamental conception of the nature of science [is the] economy of thought' and 'the goal [of physical science] is the simplest and most economical abstract expression of the facts'; see Mach, E. (1943) *Popular Scientific Lectures* fifth edition La Salle, IL: Open Court, p. 207.
12 Gerald Holton characterises Einstein's philosophical development as a pilgrimage from Mach's phenomenalistic empiricism to a more mature rationalistic realism; see Holton, G. (1973) *Thematic Origins of Scientific Thought* Cambridge, MA.: Harvard University Press. Fine, A. (1986) *The Shaky Game* Chicago: Chicago University Press argues that 'as early as 1918, Einstein's expressions of realism are presented in terms of

motivations for the pursuit of science' not so much to seek 'realist theories' but to break free from the 'chains of the "merely personal"'; see pp. 109–11. But there is little doubt that during the period up to 1920 Einstein's work was strongly influenced by Machian strictures on simplicity; see Ray, C. (1987) *The Evolution of Relativity* Bristol: Adam Hilger.

13 A complete discussion of the principles of equivalence may be found in Will, C.M. (1981) *Theory and Experiments in Gravitational Physics* Cambridge: Cambridge University Press; see also Ray, C. (1987) op. cit., ch. 2.

14 Such curvature terms represent the coupling of the law with the gravitational field; because of the emergence of these terms and other difficulties, Zahar argues that the principle of equivalence cannot strictly speaking be true of GTR, and that we must see it as a heuristic principle directing us towards the final form of the equations of GTR. See Zahar, E. (1980) 'Einstein, Meyerson, and the role of mathematics in physical discovery' *British Journal for the Philosophy of Science* **31** pp. 1–43 and Zahar, E. (1989) *Einstein's Revolution: a Study in Heuristic* La Salle, IL: Open Court.

15 Popper, K. (1963) *Conjectures and Refutations* London: Routledge.

16 Edmund Halley had considered Newton's problem about the extent of space and its contents, but suggested that Newton's resolution might have inherent difficulties: an infinite number of stars in a static universe would have resulted in a totally bright night sky – this reasoning was later developed by Olbers' in the early nineteenth century and the problem was named Olbers' Paradox. Hence, Neumann and other nineteenth-century physicists took the view that the number of stars should be finite. See Halley, E. (1720) 'Of the fixed stars' *Royal Society Transactions* **31** pp. 22–6 and, for a full discussion of Olber's Paradox, Sciama, D.W. (1961) *The Unity of the Universe* New York: Doubleday.

17 Pais, A. (1982) op. cit. gives an excellent account of the history of GTR's predictions; Will, C.M. (1988) *Was Einstein Right?* Oxford: Oxford University Press provides a contemporary account of the experimental tests of GTR.

18 Einstein's 1911 prediction appears in 'On the influence of gravitation on the propagation of light' in Einstein, A. et al. (1923) op. cit. pp. 97–1080.

19 Hawking, S.W. (1982) op. cit.

20 Einstein, A. (1917) op. cit.

21 Friedmann, A. (1922) 'On the curvature of space' *Zeitschrift fur Physik* **10** pp. 377–86 and Friedmann, A. (1924) 'On the possibility of a world with constant negative curvature' *Zeitschrift fur Physik* **21** pp. 326–32; these papers are reprinted in translation in Bernstein, J. and Feinberg, G. (eds) (1986) *Cosmological Constants: Papers in Modern Cosmology* New York: Columbia University Press.

22 Einstein, A. (1923) 'A note on the work of A. Friedmann' *Zeitschrift fur Physik* **16** p. 228.

23 Misner, C.W., Thorne, K.S., and Wheeler, J.A. (1973) *Gravitation* New York: Freeman.

24 Gödel, K. (1949) 'An example of a new type of cosmological solution' *Review of Modern Physics* **21** pp. 447–50; Oszvàth, I. and Schücking, E. (1969) 'The finite rotating universe' *Annals of Physics New York* **55** pp. 166–204.
25 Weyl, H. 'Gravitation and electricity' in Einstein, A. *et al.* (1923) op. cit., pp. 200–16.
26 Davies, P.C.W. (1982) *The Accidental Universe* Cambridge: Cambridge University Press and Hawking, S.W. (1983) 'The cosmological constant' in McCrea, W.H. and Rees, M.J. (eds) *The Constants of Physics* London: Royal Society.
27 Blau, S.K. and Guth, A.H. 'Inflationary cosmology' in Hawking, S.W. and Israel, W. (eds) (1987) *300 Years of Gravitation* Cambridge: Cambridge University Press; see also the excellent review by Abbott, L. (1988) 'The mystery of the cosmological constant' *Scientific American* **258** 5 pp. 106–13.
28 See Chapter 10 and also Ray, C. (1987) op. cit., ch. 5.
29 Cartwright, N. (1982) *How the Laws of Physics Lie* Oxford: Oxford University Press. We might try to argue that a fundamental law might be true when it belongs to a unified theory; but sadly the history of science reveals little evidence that we are converging on such a theory: if anything, the last one hundred years have shown a marked divergence as we have moved from a context with two basic forces (gravity and electromagnetism) to one with three forces (gravity, strong and electro-weak). People keep dreaming of unified theories, but nature keeps throwing up surprises. And, as Cartwright points out, although

> we want laws that unify . . . what happens may well be varied and diverse. We are lucky we can organise phenomena at all. There is no reason to think that the principles that best organize will be true, nor that the principles that are true will organize much.
>
> (Cartwright 1982 op. cit.: 53)

See Hawking, S.W. (1988) *A Brief History of Time* London: Bantam Press; and Davies, P.C.W. and Brown, J.R. (eds) (1988) *Superstrings: a Theory of Everything?* Cambridge: Cambridge University Press – note especially Sheldon Glashow's scepticism about unification and his remark, directed at superstring theory, that he is 'waiting for the superstring to break': p. 180.
30 Hawking, S.W. and Ellis, G.F.R. (1973) *The Large-Scale Structure of Space-Time* Cambridge: Cambridge University Press.
31 Barrow, J.D. and Tipler, F.J. (1986) *The Anthropic Cosmological Principle* Oxford: Oxford University Press; Hawking, S.W. (1982) op. cit.; and Davies, P.C.W. (1982) op. cit.
32 Carter, B. (1974) 'Large number coincidences and the anthropic principle in cosmology' in Longair, M.S. (ed.) *Confrontation of Cosmological Theories with Observational Data* Dordrecht: Reidel pp. 291–8.
33 Davies, P.C.W. and Brown, J.R. (eds) (1986) *The Ghost in the Atom* Cambridge: Cambridge University Press discuss a variety of views on the idea that quantum considerations show that the observer is somehow responsible for the existence of universe including Wheeler's

claim that observers are required to bring the universe into existence in line with some quantum cosmological feedback process. The link between anthropocentricity and dimensionality, mentioned briefly in the third chapter of this book, is discussed by Barrow, J.D. (1983) 'Dimensionality' in McCrea, W.H. and Rees, M.J. (eds.) (1983) op. cit., pp. 337–46.

34 The other four parameters are: the Hubble constant H, which characterises the rate at which galaxies are moving apart; the photon/proton ratio S, which plays an essential role in the formation of galaxies; the number of protons N within the observable universe; the parameter q, which describes the rate at which the expansion of the universe is slowing down. Were they all not approximately their currently observed values, then the universe would be inconsistent with the possibility of life; see Barrow, J.D. and Tipler, F.J. (1986) op. cit. for a full discussion. Rees, M.J. (1983) 'Large numbers in astrophysics', in McCrea, W.H. and Rees, M.J. (eds) (1983) op. cit., pp. 311–22, carries the discussion further with a review of the 'large number' coincidences in cosmology.

35 Davies, P.C.W. (1982) op. cit., p. 121.

36 Earman, J.S. (1987) 'The SAP also rises: a critical examination of the anthropic principle' *American Philosophical Quarterly* **24** pp. 307–17.

37 See Chapter 10 for a discussion of inflationary cosmology. Note also the discussion of SAP and cosmological perspectives in Hacking, I. (1987) 'The Inverse Gambler's fallacy; the argument from design. The anthropic principle applied to Wheeler universes' *Mind* **96** pp. 331–40.

38 Hawking, S.W. (1982) op. cit., p. 425.

39 Carter, B. (1974) op. cit.

40 Carter, B. (1983) 'The anthropic principle and its implications for biological evolution' in McCrea, W.H. and Rees, M.J. (eds) (1983) op. cit., pp. 347–63.

10 COSMOLOGICAL CONUNDRUMS

1 Leibniz's argument is almost exactly that given (earlier) by the English philosopher Thomas Hobbes (1588–1679) in his essay 'On liberty and necessity' – see the reference to Hobbes' argument in Anscombe, G.E.M. (1986) 'Times, beginnings, and causes' in Kenny, A. (ed.) *Rationalism, Empiricism, and Idealism* Oxford: Oxford University Press p. 93. For the quotation from Leibniz see Alexander, H.G. (ed.) (1956) *The Leibniz–Clarke Correspondence* Manchester: Manchester University Press pp. 36–8.

2 Presumably a reference to Descartes's views.

3 See Alexander, H.G. (ed.) (1956) op. cit. p. 49.

4 Details of the initial discovery are given in Penzias, A.A. and Wilson, R.W. (1965) 'A measurement of excess antenna temperature at 4080 Mc/s' *Astrophysical Journal* **142** pp. 419–21; a full discussion of its implications may be found in Raychaudhuri, A.K. (1979) *Theoretical Cosmology* Oxford: Oxford University Press ch. 6.

5 The proofs that singularities must exist in GTR were provided by

Roger Penrose and Stephen Hawking in the 1960s. See Penrose, R. (1965) 'Gravitational collapse and spacetime singularities' *Physical Review Letters* **14** pp. 57–9; and Hawking, S.W. (1966) 'The occurrence of singularities in cosmology' *Proceedings of the Royal Society* **A 294** pp. 511–21.

6 See Geroch, R.P. and Horowitz, G.T. (1979) 'Global structures of spacetimes' in Hawking, S.W. and Israel, W. (eds) *General Relativity* Cambridge: Cambridge University Press.

7 See the original idea in Bondi, H. and Gold, T. (1948) 'The steady-state theory of the expanding universe' *Monthly Notices of the Royal Astronomical Society* **108** pp. 252–70 and Hoyle, F. (1948) 'A new model for the expanding universe' *Monthly Notices of the Royal Astronomical Society* **108** pp. 372–82; and also the (supportive) comments on the theory by Narlikar, J.V. (1988) *The Primeval Universe* Oxford: Oxford University Press pp. 217–28. For a less partisan account see Barrow, J.D. (1988) *The World within the World* Oxford: Oxford University Press pp. 212–18.

8 This point is made by many who refuse to grant any force to the 'cosmological argument' for the existence of God; this argument starts from the premiss that nothing happens without a cause (or reason), and leads us to the conclusion that God exists on the grounds that there must be some cause (or reason) for the existence of the world as a whole; see, for example, the text of a radio debate between Bertrand Russell and Frederick Copleston in Vesey, G. (ed.) (1974) *Philosophy in the Open* Milton Keynes: Open University Press pp. 115–20.

9 Hawking, S.W. (1987) 'Quantum cosmology' in Hawking, S.W. and Israel, W. (eds) *300 Years of Gravitation* Cambridge: Cambridge University Press; see also the introduction to Hawking, S.W. and Israel, W. (eds) (1979) op. cit.

10 See Hawking, S.W. (1987) op. cit., especially pp. 633–6.

11 Hawking, S.W. (1987) op. cit., p. 635; see also the comments on this proposal by Grunbaum, A. (1989) 'The pseudo-problem of creation in physical cosmology' *Philosophy of Science* **56** p. 393; this article by Grunbaum is reprinted in the valuable collection by Leslie, J. (ed.) (1989) *Physical Cosmology and Philosophy* New York: Macmillan.

12 Similarly, if the universe does collapse, there may be no definite end-point to the future. But Hawking argues that even universes which are non-singular at the start may recollapse to a singularity at the end. See Hawking, S.W. (1987) op. cit., p. 650.

13 There are several excellent general reviews of the important features of inflationary cosmology; non-technical accounts of the original theory may be found in Guth, A.H. and Steinhardt, P.J. (1984) 'The inflationary universe' *Scientific American* May 1984, reprinted in Davies, P.C.W. (ed.) (1989) *The New Physics* Cambridge: Cambridge University Press; Narlikar, J.V. (1988) op. cit., ch. 5; and (in brief) in Gribbin, J. and Rees, M. (1989) *Cosmic Coincidences* London: Bantam Press pp. 277–83; a comprehensive but technical account of the original and revised theories is given in Blau, S.K. and Guth, A.H. (1987) 'Inflationary cosmology' in Hawking, S.W. and Israel, W. (1987) (eds)

op. cit., pp. 524–604. A comprehensive non-technical account of the refined theory is given in Guth, A.H. (1989) 'Starting the universe' in Cornell, J. (ed.) *Bubbles, Voids and Bumps in Time: the New Cosmology* Cambridge: Cambridge University Press.

14 Ellis, G.F.R. and Williams, R.M. (1988) *Flat and Curved Space-Times* Oxford: Oxford University Press pp. 277–85 gives a more detailed discussion of this problem.
15 Blau, S.K. and Guth, A.H. (1987) op. cit., p. 542.
16 Guth, A.H. (1989) op. cit., p. 136; note that zero degrees Kelvin equals roughly minus 273 degrees Celsius or centigrade.
17 Without any recognisable, distinctive particles and antiparticles, the idea of such a high temperature is spelled out in terms of excitations in the matter field instead of the more usual terms of average particle speeds.
18 Blau, S.K. and Guth, A.H. (1987) op. cit.
19 Blau, S.K. and Guth, A.H. (1987) op. cit., p. 541.
20 See the discussion of this analogy in Hawking, S.W. (1988) *A Brief History of Time* London: Bantam Press pp. 127–8.
21 Guth, A.H. and Steinhardt, P.J. (1984) op. cit.
22 This point is made with some force by Grunbaum, A. (1989) op. cit., pp. 389–93, who is reacting to a claim, by the astronomer Bernard Lovell, that inflationary theory allows us to say that the universe was created out of nothing.
23 Hawking, S.W. (1988) op. cit., p. 128.
24 Linde, A. (1985) 'The universe: inflation out of chaos' *New Scientist* **105**, 1446 pp. 14–18; reprinted in Leslie, J. (ed.) (1989) op. cit.
25 A translation of Laplace's (1799) paper discussing the possibility of 'black holes' appears as Appendix A in Hawking, S.W. and Ellis, G.F.R. (1973) *The Large Scale Structure of Space-Time* Cambridge: Cambridge University Press. Paul Davies, in an article in the *Sunday Correspondent*, April 1990, notes that the earliest mention of the idea of a black hole may have been in 1783, by the clergyman John Michell, who addressed himself to the same problems in Newtonian theory as Laplace.
26 See Chandrasekhar's remarks on this incident in Mehra, J. (ed.) (1973) *The Physicist's Conception of Nature* Dordrecht: Reidel, pp. 36–7.
27 Hawking, S.W. and Israel, W. (eds) (1979) op. cit., see their 'Introduction'.
28 Penrose, R. (1979) 'Singularities and time-asymmetry' in Hawking, S.W. and Israel, W. (eds) (1979) op. cit., p. 618.
29 See Earman, J.S. (1971) 'Laplacian determinism, or is this any way to run a universe?' *Journal of Philosophy* **68** pp. 729–44; and Malament, D.B. (1977) 'Observationally indistinguishable spacetimes' in Earman, J.S., Glymour, C.N., and Stachel, J.J. (eds) *Foundations of Space-time Theories* Minneapolis: University of Minnesota Press.
30 See Geroch, R.P. and Horowitz, G.T. (1979) op. cit.
31 See Clarke, C.J.S. (1976) 'Spacetime singularities' *Communications in Mathematical Physics* **49** pp. 17–23; and Clarke, C.J.S. (1982) 'Singular spacetimes' *Communications in Mathematical Physics* **84** pp. 329–31. See

also the comments by Earman, J.S. (1989) *World Enough and Space-Time* Cambridge, MA: MIT Press, chs 5, 8 and 9.
32 See Ray, C. (1987) *The Evolution of Relativity* Brisol: Adam Hilger, ch. 4; and Earman, J.S. (1989) op. cit., chs 8 and 9.
33 For a more detailed general review of the philosophical issues connected with black holes and cosmic censorship see Weingard, R. (1979) 'Some philosophical aspects of black holes' *Synthèse* **42** pp. 191–219. An excellent general discussion of the physics of black holes appears in Wald, R.M. (1977) *Space, Time, and Gravity* Chicago: Chicago University Press.

CONCLUSION: RELATIVITY – JUST ANOTHER BRICK IN THE WALL?

1 Hawking, S.W. (1980) *Is the End of Theoretical Physics in Sight?* Cambridge: Cambridge University Press; this is a reprint of Hawking's lecture on being made Lucasian Professor of Mathematics at Cambridge University.
2 Cartwright, N. (1982) *How the Laws of Physics Lie* Oxford: Oxford University Press; Hacking, I. (1983) *Representing and Intervening* Cambridge: Cambridge University Press; and Galison, P. (1987) *How Experiments End* Chicago: Chicago University Press.
3 Kuhn, T.S. (1970) *The Structure of Scientific Revolutions* second edition Chicago: Chicago University Press: see postscript; and Lakatos, I. (1970) 'Falsification and the methodology of scientific research programmes' in Lakatos, I. and Musgrave, A. (eds) *Criticism and the Growth of Knowledge* Cambridge: Cambridge University Press.
4 Kuhn, T.S. (1970) op. cit., p. 182.
5 Kuhn, T.S. (1970) op. cit. and Kuhn, T.S. (1974) 'Second thoughts on paradigms' in Suppe, F. (ed.) *The Structure of Scientific Theories* Urbana, IL: University of Illinois Press; Hesse, M.B. (1974) *The Structure of Scientific Inference* London: Macmillan and Hesse, M.B. (1980) *Revolutions and Reconstructions* Bloomington, IN: Indiana University Press; Lakatos, I. (1970) op. cit.; and Laudan, L. (1986) *Science and Values* Berkeley: University of California Press.
6 Kuhn, T.S. (1970) op. cit., pp. 181–7.
7 Galison, P. (1988) 'Philosophy in the laboratory' *Journal of Philosophy* **85** p. 10.
8 See Davies, P.C.W. and Brown, J. (eds) (1988) *Superstrings: a Theory of Everything?* Cambridge: Cambridge University Press; for more technical detail see the opening review of string theory in Green, M.B., Schwarz, J.H., and Witten, E. (1987) *Superstring Theory* vol. 1 Cambridge: Cambridge University Press.
9 See Blau, S.K. and Guth, A.H. (1987) 'Inflationary cosmology' in Hawking, S.W. and Israel, W. (eds) *300 Years of Gravitation* Cambridge: Cambridge University Press pp. 524–603.
10 Wald, R. M. (1984) *General Relativity* Chicago: Chicago University Press; see pp. 450–69.
11 Damour, T. (1987) 'The problem of motion in Newtonian and

Einsteinian gravity' in Hawking, S.W. and Israel, W. (eds) (1987) op. cit., pp. 128–98.
12 Zahar, E. (1989) *Einstein's Revolution: a Study in Heuristic* La Salle, IL: Open Court.
13 See the interesting essay by Kuhn, T.S. (1981) 'A function for thought experiments' in Hacking, I. (ed.) *Scientific Revolutions* Oxford: Oxford University Press.
14 See Laudan, L. (1986) op. cit., and also Hesse, M.B. (1980) op. cit. for investigations into the role of values in science. See also the 'sociologically' based analysis of those like Bloor, Latour, and Woolgar, who argue that scientific knowledge is a function of sociology – that science is no more than a social construction. This programme is an extreme extension of Kuhn's thinking. There is a particular focus in these accounts on the way that scientists, with all their petty attitudes and rivalries and values and so on, and with all the constraints which society imposes on the scientific community, literally negotiate amongst themselves what ideas should emerge from their research – it is as if they can say what they will about the world. This strikes me as a curious attitude. Yes, science is undoubtedly influenced by society. But it is also influenced by the way the world is. For details of those who pursue the sociology of science, see Richards, S. (1987) *Philosophy and Sociology of Science* second edition Oxford: Basil Blackwell.
15 See, for a discussion of the importance of symmetries and invariance for physics, van Fraassen, B. (1989) *Laws and Symmetries* Oxford: Oxford University Press.
16 Cartwright, N. (1982) op. cit.

SELECT BIBLIOGRAPHY

FIVE KEY BOOKS FOR FURTHER STUDY

Alexander, H.G. (ed.) (1956) *The Leibniz–Clarke Correspondence* Manchester: Manchester University Press

Flood, R. and Lockwood, M. (eds) *The Nature of Time* Oxford: Basil Blackwell

Pais, A. (1982) *Subtle is the Lord* Oxford: Oxford University Press (a brilliant biography of Einstein)

Reichenbach, H. (1957) *The Philosophy of Space and Time* New York: Dover

Sklar, L. (1974) *Space, Time, and Spacetime* Berkeley, CA: University of California Press

TEN USEFUL BOOKS WITH A PHILOSOPHICAL SLANT

Angel, R.B. (1980) *Relativity: the Theory and Its Philosophy* Oxford: Pergamon

Earman, J.S. (1989) *World Enough and Space-Time* Cambridge, MA: MIT Press

van Fraassen, B.C. (1980) *An Introduction to the Philosophy of Time and Space* second edition New York: Random House

Friedman, M. (1983) *Foundations of Space-Time Theories* Princeton, NJ: Princeton University Press

Horwich, P. (1987) *Asymmetries in Time* Cambridge, MA: MIT Press

Mellor, D.H. (1981) *Real Time* Cambridge: Cambridge University Press

Munitz, M.K. (1986) *Cosmic Understanding: Philosophy and the Science of the Universe* Princeton, NJ: Princeton University Press

Newton-Smith, W.H. (1980) *The Structure of Time* London: Routledge

Ray, C. (1987) *The Evolution of Relativity* Bristol: Adam Hilger

Salmon, W.C. (1980) *Space, Time, and Motion* second edition Minneapolis: University of Minnesota Press

TEN USEFUL BOOKS WITH A SCIENTIFIC OR MATHEMATICAL SLANT

Barbour, J.B. (1989) *Absolute or Relative Motion?* vol. I Cambridge: Cambridge University Press (look out for vol. II)

SELECT BIBLIOGRAPHY

Cohen, I.B. (1987) *The Birth of the New Physics* Oxford: Oxford University Press
Eddington A.S. (1920) *Space, Time, and Gravitation* Cambridge: Cambridge University Press
Ellis, G.F.R. and Williams, R.M. (1988) *Flat and Curved Space-Times* Oxford: Oxford University Press
French, A.P. (1968) *Special Relativity* London: Van Nostrand Reinhold
Gribbin, J. and Rees, M. (1989) *Cosmic Coincidences* London: Bantam
Hawking, S.W. (1988) *A Brief History of Time* London: Bantam
Stewart, I. (1987) *The Problems of Mathematics* Oxford: Oxford University Press
Toretti, R. (1983) *Relativity and Geometry* Oxford: Pergamon
Wald, R.M. (1984) *General Relativity* Chicago: Chicago University Press

TEN USEFUL COLLECTIONS OF ARTICLES AND CLASSIC PAPERS

Bernstein, J. and Feinberg, G. (eds) (1986) *Cosmological Constants: Papers in Modern Cosmology* New York: Columbia University Press
Capek, M. (ed.) (1976) *The Concepts of Space and Time* Dordrecht: Reidel
Cornell, J. (ed.) (1989) *Bubbles, Voids and Bumps in Time: the New Cosmology* Cambridge: Cambridge University Press
Earman, J.S., Glymour, C.N., and Stachel, J.J. (eds) (1977) *Foundations of Space-Time Theories* Minneapolis: University of Minnesota Press
Einstein, A., Lorentz, H.A., Weyl, H., and Minkowski, H. (1923) *The Principle of Relativity* London: Methuen
Hawking, S.W. and Israel, W. (eds) (1987) *300 Years of Gravitation* Cambridge: Cambridge University Press
Healey, R.A. (ed.) (1981) *Reduction, Time, and Reality* Cambridge: Cambridge University Press
Leslie, J. (ed.) (1989) *Physical Cosmology and Philosophy* New York: Macmillan
Salmon, W.C. (ed.) (1970) *Zeno's Paradoxes* Indianapolis, IN: Bobbs-Merrill
Schilpp, P. (ed.) (1969) *Albert Einstein: Philosopher-Scientist* La Salle, IL: Open Court.

FIVE USEFUL INTRODUCTIONS TO PROBLEMS IN THE PHILOSOPHY OF SCIENCE

Chalmers, A.F. (1982) *What is This Thing Called Science?* second edition Milton Keynes: Open University Press
Hacking, I. (1983) *Representing and Intervening* Cambridge: Cambridge University Press
Kuhn, T.S. (1970) *The Structure of Scientific Revolutions* second edition Chicago: Chicago University Press
Newton-Smith, W.H. (1981) *The Rationality of Science* London: Routledge
O'Hear, A. (1989) *An Introduction to the Philosophy of Science* Oxford: Oxford University Press

SELECT BIBLIOGRAPHY

The literature on space and time contains many fine books in addition to the ones mentioned above. Further references may be found in the notes to each chapter. Inevitably many have not been mentioned in this personal selection. But I believe that you will find every book in this bibliography interesting and rewarding, though some will be hard going. Unfortunately, not every book mentioned above is still in print; however, inter-library loan facilities may help in locating copies, should any be hard to find.

INDEX

absolute space 99–103, 134–7, 146–50, 179–84, 215–16
absolute spacetime 134–9, 146–50, 215–16
absolute time 102
Achinstein, P. 242n.
Adams, J.C. 90
ad hoc hypotheses 91, 184
affine geometry 54, 134, 139–42, 197
Alexander, H.G. 241n., 244n., 245n., 246n., 255n.
Anscombe, G.E.M. 255n.
Anthropic Principle: 83, 226–7; Strong 189–92, 222; Weak 191–2
Aristotle 5, 109, 229n., 249n.
Armstrong, D.M. 241n.
Asquith, P.A. 239n., 251n.
Ayer, A.J. 241n.

backwards causation 153, 171–5
Barbour, J.B. 240n., 241n., 244n., 248n.
Barnes, J. 229n.
Barrow, J.D. 83, 238n., 254n., 255n.
Bastin, E. 229n.
Beck, A. 230n.
beginning of time 199–203
Beltrami, E. 71
Benacerraf, P. 14, 230n.
Bentley, R. 177
Bergson, H. 25–6, 44–5, 230n.
Berkeley, G. 116, 122, 126, 243n.
Bernstein, J. 248n., 253n.

Berresford, G.C. 230n.
Bertotti, B. 244n., 248n.
big bang 84–6, 146, 196–9, 204–9
black holes 195, 201–3, 209–15, 224
Black, M. 20, 239n.
Blau, S.K. 206, 254n., 256n., 257n., 258n.
Bloor, D. 259n.
Bohm, D. 42, 61, 233n., 235n.
Bolyai, J. 71
Bondi, H. 145–6, 200, 232n., 249n., 256n.
Bowman, P.A. 57, 234n.
Boyle, R. 104, 113
Brans, C. 244n., 248n.
Brown, H.R. 234n.
Brown, J. 238n., 254n., 258n.
Butterfield, J. 249n.

Calder, N. 236n.
Capek, M. 230n.
Carnap, R. 244n.
Cartan's spacetime reformulation of Newtonian gravity 247n.
Carter, B. 189, 192, 254n., 255n.
Cartwright, N. 188, 218, 226–7, 254n., 258n., 259n.
Cauchy, A. 11, 212–13, 216, 229n., 249n.
Chandler, M. 237n.
Chandrasekhar 210, 257n.
Chiu, H.Y. 247n.
Clarke, C.J.S. 213, 257n.
Clarke, S. 99–100, 103, 106–12, 194, 207, 242n.

INDEX

Cleomedes 242n.
clock paradox 24–6
closed causal loops 79–82, 153–71
conformal geometry 53–4, 97
conventionalism 48, 90–8
Cornell, J. 257n.
cosmic censorship 195, 211–15
cosmological constant 176–92, 223
cosmological models 84–90, 133, 143, 186, 196–209
Cosmological Principle 86–90
Cottingham, J. 234n.
covariance 147

Damour, T. 258n.
Davies, P.C.W. 190–1, 238n., 250n., 254n., 256n., 257n., 258n.
Descartes, R. 104, 113, 240n., 241n., 255n.
determinism 148–50, 215–16
Devitt, M. 234n.
Dicke, R. 244n., 248n.
dimensionality 5, 82–3
Doppler effect 45
Duff, M.J. 252n.
Duhem, P. 90, 230n., 239n.
Duhem–Quine thesis 22, 90, 92
Dummett, M. 156–7, 250n., 251n.

Earman, J.S. 115, 146, 148–50, 190, 216, 235n., 241n., 242n., 243n., 246n., 247n., 249n., 251n., 255n., 257n., 258n.
Eddington, A.S. 1, 74, 210, 236n.
Ehrenfest, P. 82, 238n.
Einstein, A. 1–4, 24–5, 82, 119, 131–3, 139, 146–8, 176–7, 179–87, 221–3, 225, 230n., 236n., 243n., 244n., 247n., 249n., 252n., 253n., 254n.
Einstein's static universe 180–4
Elliot, R. 66–8, 235n.
Ellis, B. 57, 234n., 235n.
Ellis, G.F.R. 88–9, 145, 188, 233n., 235n., 238n., 239n., 248n., 249n., 250n., 254n., 257n.
empirical equivalence 92–5
Euclid 69, 70
Euclidean geometry 69–78

expansion of spacetime 84–90, 196–9, 204–9
experimentation 127–30, 224–5

Feigl, H. 245n.
Feinberg, G. 248n., 253n.
Feyerabend, P. 126, 128, 245n.
Feynman, R. 64–5, 83, 91, 164–5, 235n., 238n., 239n., 250n.
Feynman diagrams 64–5, 164–5
field equations of GTR 78, 131–2, 139–46, 176–89
Fine, A. 240n., 252n.
Fizeau, A. 46
Flood, R. 250n.
Foucault, J. 46
French, A.P. 42, 230n., 232n., 233n.
Friedman, M. 97, 114–15, 122, 234n., 235n., 237n., 240n., 243n., 244n., 246n., 247n., 248n., 249n.
Friedmann, A. 85–6, 133, 143–4, 186, 196, 212, 225, 248n., 253n.
Frisch, D.H. 230n.

Galileo 120
Galison, P. 218, 220–1, 245n., 258n.
Gamov, G. 251n.
Gardner, M.R. 246n.
Gauss, C.F. 71
Geroch, R. 210, 256n., 257n.
Giannoni, C. 234n.
Giere, R. 239n.
Gilman, R.C. 244n., 248n.
Glashow, S. 254n.
Glymour, C.N. 246n., 249n., 257n.
Gödel, K. 142, 186, 248n., 250n., 251n., 253n.
Gold, T. 200, 256n.
Graves, J.C. 243n.
Green, M.B. 70, 82–3, 224, 258n.
Greenstreet, W.J. 236n.
Gribbin, J. 238n., 243n., 256n.
Grunbaum, A. 52–55, 59, 234n., 235n., 256n.
Guth, A. 204–8, 254n., 256n., 257n., 258n.

INDEX

Hacking, I. 127–30, 218, 245n., 259n.
Halfele–Keating experiment 26, 231n.
Hall, A.R. 241n., 243n.
Hall, M.B. 241n., 243n.
Halley, E. 253n.
Hanfling, O. 244n.
Hanson, N.R. 95, 240n., 244n.
Harré, R. 233n.
Harrison, J. 168, 251n.
Hartle, J. 202
Hawking, S.W. 145, 176, 188, 191, 201–2, 208, 210, 217, 221, 224, 229n., 235n., 236n., 238n., 248n., 249n., 250n., 253n., 254n., 255n., 256n., 257n., 258n.
Healey, R.A. 243n., 244n.
Hesse, M.B. 219, 230n., 239n., 240n., 244n., 258n., 259n.
Hobbes, T. 255n.
Hoefer, C. 248n.
Hoffman, B. 233n., 234n.
Hoffman, W. 247n.
hole argument 146–50, 213–16
Holton, G. 252n.
homogeneity 86–90
Hookway, C. 230n.
Horowitz, G.T. 256n., 257n.
Horwich, P. 170, 250n., 251n.
Hoyle, F. 200, 256n.
Hubble, E. 146, 184
Hume, D. 122
Huygens, C. 99, 110–11, 246n.

identity of indiscernibles 67, 107–8, 148
Iltis, C. 243n.
indeterminism 148–50, 215–16
inertial forces 27–32, 99–103, 113–15, 118–20
inertial frames 27–32
inertial motion 27–32, 99–103, 118–20, 134–9
infinitesimals 10
infinite speed 66–8
inflationary cosmology 70–1, 83, 195, 203–9
instrumentalism 150

intervals, spacetime 36–7
invariance 36, 61, 226
Isham, C.J. 252n.
isotropy 86–90
Israel, W. 211, 229n., 236n., 238n., 250n., 254n., 255n., 256n., 257n., 258n.

Kagon, R. 242n.
Kenny, A. 255n.
Kerr metric: rotating black hole 142, 248n.
Kerszberg, P. 252n.
Kitcher, P. 251n.
Koyré, A. 241n.
Kretschmann, E. 249n.
Kuhn, T.S. 95, 124–5, 219, 220, 225–6, 240n., 245n., 258n., 259n.

Lakatos, I. 219–20, 226, 258n.
Laplace, P.S. 209, 212, 257n.
Latour, B. 259n.
Laudan, L. 219, 258n.
laws of physics 187–9, 221, 226–7
Leibniz, G. 3, 67, 99–100, 105–13, 118, 131, 148, 193–5, 203, 241n., 242n., 243n., 244n., 245n., 246n., 255n.
Leibniz–Clarke correspondence 108–13, 193–5, 203
Leplin, J. 233n., 245n.
Leslie, J. 256n., 257n.
Le Verrier, J. 90–1, 185
Lewis, D. 251n.
light cone 60
limiting values 11–14
Linde, A. 70, 83, 204, 209, 236n., 238n., 257n.
Lobatschewskii, N.I. 71
Locke, J. 104–5, 113, 223, 241n.
Lockwood, M. 250n.
Longair, M.S. 254n.
Lorentz, H.A. 58, 131–2, 230n., 236n., 252n.
Lorentz transformations 27, 33–6

McCrea, W.H. 238n., 254n., 255n.
Mach, E. 2–3, 102, 115–23, 125, 130–1, 133–4, 138, 144–6, 177,

179–83, 222, 227, 243n., 244n., 247n., 252n.
Machamer, P.K. 243n.
Mach's Principle 133–4, 138–9, 143–6, 181–4, 222
Malament, D.B. 53, 97, 234n., 247n., 251n.
Manders, K. 243n.
manifold, spacetime 54, 131, 148–9
Marder, L. 232n.
Mates, B. 241n.
Maudlin, T. 249n.
Maxwell, G. 245n.
Maxwell, J.C. 24
Mehra, J. 257n.
Mellor, D.H. 152, 165, 172–4, 249n., 250n., 251n.
metrical geometry 54, 131, 139, 140–2
Meyerson, E. 253n.
Michell, J. 257n.
microwave background 196
Minkowski, H. 73, 97, 132, 138, 140, 144, 212, 230n., 247n., 250n., 252n.
Misner, C. 246n., 247n., 248n., 249n., 253n.
Mundy, B. 243n.
Munitz, M. 251n.
Murdoch, D. 234n.
Musgrave, A. 258n.

naked singularities 214
Narlikar, J.V. 256n.
Nerlich, G. 235n., 237n.
Neumann, C. 177–8, 180, 182–3, 252n.
Newton, I. 2, 34, 99–105, 107, 109, 110–12, 117–20, 123, 132, 135–7, 180–2, 184–5, 203, 240n., 241n., 242n., 243n., 246n., 247n., 252n.
Newton-Smith, W.H. 22, 230n., 232n., 233n., 238n., 244n., 245n.
Newton's thought experiments 100–3, 110–12, 118–20, 133, 225
non-Euclidean geometry 2, 69–78
North, J.D. 252n.
Norton, J. 146, 249n., 253n.

Occam's razor 97, 116, 122, 240n.
Occam, William of 240n.
Olber's paradox 253n.
Oszvàth, I. 142, 144, 186, 254n.
Owen, G.E.L. 10, 229n.

Pais, A. 176, 181, 184, 229n., 249n., 251n., 252n., 253n.
Papineau, D. 240n.
paradoxes: clock 24–6; lamp super-task 14–15; parallel super-task 15–19; time travel 166–71; triplets 41–4; twins 36–41, 134; Zeno's 5–11.
Parfit, D. 235n.
Peirce, C.S. 229n.
Penrose, R. 6, 150, 155, 210–11, 213, 221, 229n., 238n., 250n., 255n., 257n.
Penzias, A. 196, 255n.
Planck, H. 223
Poincaré, H. 70–1, 75, 79, 90, 92, 235n., 236n.
Poisson's equation 178
Popper, K.R. 91, 183, 239n., 253n.
positivism 40, 117, 122–30, 190–1, 222
primary qualities 104
Putnam, H. 245n.

Quine, W.V.O. 90–1, 234n., 239n.

Raine, D. 143–4, 244n., 248n.
Ray, C. 235n., 240n., 241n., 244n., 248n., 251n., 253n., 254n., 258n.
Raychaudhuri, A.K. 255n.
realism 48
Recami, E. 66, 235n.
Redmount, I. 250n.
Rees, M. 238n., 254n., 255n., 256n.
Reichenbach, H. 52, 70–1, 75, 78–9, 80–1, 91, 95–6, 131–2, 234n., 236n., 237n., 238n., 239n., 244n., 246n.
Reichenbach, M. 236n.
relationism 105–8, 118–20, 138–9, 215–16
Richards, S. 259n.
Riemann, G. 72, 82, 236n.

INDEX

Riggs, P.J. 251n.
Rindler, W. 248n.
Rømer, O. 57–8, 234n.
Russell, B. 256n.

Sainsbury, M. 14–17, 19, 29, 230n.
Salmon, W. 6, 20, 46, 49, 50, 53–4, 229n., 230n., 233n., 234n., 235n.
Schilpp, P. 244n.
Schücking, E. 142, 144, 186, 254n.
Schwarz, J.H. 70, 82–3, 224, 236n., 238n., 258n.
Schwarzschild solution, the 132, 141, 144, 222
Sciama, D.W. 143, 244n., 248n.
secondary qualities 104
Seeliger, H. 177–8, 180, 182–3
simplicity 120–2, 222
Simplicius 5, 10, 229n.
simultaneity: absolute 53–7, 136; metrical 51–3, 57, 97, 147; topological 51–3, 57
singularities 193–216
Sitter, W. de 133, 176, 182, 200, 251n., 252n.
Sklar, L. 103, 110, 113–15, 149, 215, 236n., 238n., 240n., 242n., 243n., 246n., 247n., 248n., 251n.
slow clock transport 57–9
Smith, J.H. 230n.
Sorabji, R. 242n.
spacetime: diagrams 27–33; expansion of 84–90, 196–9, 204–9
Stachel, J.J. 230n., 242n., 246n., 257n.
standard signal synchrony 47–56
Stein, H. 241n., 242n., 243n., 246n., 247n.
Steinhardt, P.J. 204, 256n., 257n.
Stewart, I. 10, 229n.
Stoothoff, R. 234n.
substantivalism 103, 146–50, 215–16
Sufficient Reason, Principle of 107–8, 193–5
superstrings 70–1, 82–3, 224
super-tasks 14–19
Suppe, F. 258n.
Swinburne, R. 250n.

tachyons 60–6, 153
Taylor, E.F. 232n.
Teller, P. 146, 148, 241n., 242n., 249n.
tensors 72, 147
theories 218–28
theory-dependence of observation 124–30
Thomson, J. 6, 14–15, 17, 19, 229n.
Thorne, K. 246n., 247n., 248n., 249n., 253n.
Thurston, W.P. 239n.
time, beginning of 199–203
time machines 151–71
time travel 64–6, 151–75
Tipler, F.J. 254n., 255n.
Tolman, R.C. 61, 235n.
topological holes 198–9, 213–16
topology 20–3, 79–90, 139, 141, 146–51, 154–6
Turnbull, R.G. 243n.
two-slit experiment 49–50

underdetermination of theory by data 49, 90–8

van Fraassen, B. 126–7, 129, 236n., 245n., 259n.
Vesey, G. 256n.
von Leyden, W. 241n.

Wald, R.M. 238n., 258n.
Waylen, P.C. 244n., 248n.
Weekes, J.R. 239n.
Weierstrass, K. 11
Weinberg, S. 249n.
Weingard, R. 258n.
Wells, H.G. 151, 249n.
Weyl, H. 82, 143, 186, 230n., 236n., 238n., 252n.
Wheeler, J.A. 73, 134, 209, 232n., 246n., 247n., 248n., 249n., 253n., 255n.
Williams, R.M. 88–9, 233n., 238n., 257n.
Winnie, J. 53, 234n.
Wilson, M. 243n.
Wilson, R.W. 196, 255n.
Winch, P. 241n.

INDEX

Witten, E. 258n.
Wittgenstein, L. 124
Woolgar, S. 259n.
wormholes 160–1

Young, T. 49–50

Zahar, E. 225, 245n., 253n., 259n.
Zeno 2, 5–12, 16, 20–2, 79, 229n.
Zeno's paradoxes 5, 7–11